Lecture Notes
in Business Information Processing 27

Series Editors

Wil van der Aalst
Eindhoven Technical University, The Netherlands

John Mylopoulos
University of Trento, Italy

Norman M. Sadeh
Carnegie Mellon University, Pittsburgh, PA, USA

Michael J. Shaw
University of Illinois, Urbana-Champaign, IL, USA

Clemens Szyperski
Microsoft Research, Redmond, WA, USA

Malu Castellanos   Umesh Dayal
Timos Sellis (Eds.)

Business Intelligence for the Real-Time Enterprise

Second International Workshop, BIRTE 2008
Auckland, New Zealand, August 24, 2008
Revised Selected Papers

 Springer

Volume Editors

Malu Castellanos
Umesh Dayal
Hewlett-Packard
1501 Page Mill rd, MS-1142, Palo Alto, CA 94304, USA
E-mail: {malu.castellanos,umeshwar.dayal}@hp.com

Timos Sellis
Institute for the Management of Information Systems
17 Mpakou str, Athens 11524, Greece
E-mail: timos@imis.athena-innovation.gr

Library of Congress Control Number: 2009931688

ACM Computing Classification (1998): H.3, H.2, J.1

ISSN 1865-1348

ISBN 978-3-642-03421-3 Springer Berlin Heidelberg New York

springer.com

© Springer-Verlag Berlin Heidelberg 2009

Typesetting: Camera-ready by author, data conversion by Scientific Publishing Services, Chennai, India
Printed on acid-free paper   SPIN: 12718801   06/3180   5 4 3 2 1 0

Preface

In today's competitive and highly dynamic environment, analyzing data to understand how the business is performing, to predict outcomes and trends, and to improve the effectiveness of business processes underlying business operations has become critical. The traditional approach to reporting is no longer adequate, users now demand easy-to-use intelligent platforms and applications capable of analyzing real-time business data to provide insight and actionable information at the right time. The end goal is to improve the enterprise performance by better and timelier decision making, enabled by the availability of up-to-date, high-quality information.

As a response, the notion of "real-time enterprise" has emerged and is beginning to be recognized in the industry. Gartner defines it as "using up-to-date information, getting rid of delays, and using speed for competitive advantage is what the real-time enterprise is all about... Indeed, the goal of the real-time enterprise is to act on events as they happen." Although there has been progress in this direction and many companies are introducing products toward making this vision a reality, there is still a long way to go. In particular, the whole lifecycle of business intelligence requires new techniques and methodologies capable of dealing with the new requirements imposed by the real-time enterprise. From the capturing of real-time business performance data to the injection of actionable information back into business processes, all the stages of the business intelligence (BI) cycle call for new algorithms and paradigms as the basis of new functionalities including dynamic integration of real-time data feeds from operational sources, evolution of ETL transformations and analytical models, and dynamic generation of adaptive real-time dashboards, just to name a few.

The series of BIRTE workshops aims to provide a forum for discussing topics related to this emerging field and setting research directions of business intelligence (BI) toward the vision of the real-time enterprise. Following the success of BIRTE 2006 held in Seoul, Korea in conjunction with VLDB 2006, BIRTE 2008 was held in Auckland, New Zealand, on August 24, 2008, in conjunction with VLDB 2008. It included one keynote talk and three sessions where ten papers were presented. In contrast to BIRTE 2006, on this occasion we had six invited talks given by well-known researchers from academia and industry driving major efforts in areas that are fundamental to BIRTE. The papers by the keynote speaker, four invited talks and the four accepted papers are included here.

Volker Markl (Technische Universität Berlin) gave the keynote talk on "Situational Business Intelligence." Volker is an expert in BI and has a long research record in the area. He presented the state of the art for situational applications and the impact of Web 2.0 on these applications; he also presented examples and research challenges that the information management research community needs to address in order to arrive at a platform for situational business intelligence.

We wish to express special thanks to the Program Committee members for providing their technical expertise in reviewing the submitted papers and helping us prepare an interesting program. To our keynote speaker and the presenters of the papers we express our appreciation for sharing their work and experiences in this workshop. We thank the VLDB 2008 organizers for their help and organizational support. Finally, we would like to extend many thanks to Alkis Simitsis for maintaining the workshop's website, for preparing the e-proceedings and for his help in producing this volume.

April 2009   Malu Castellanos
                                                                     Umesh Dayal
                                                                     Timos Sellis

Organization

Organizing Committee

General Chair

Umeshwar Dayal   Hewlett-Packard, USA

Program Committee Chairs

Malu Castellanos   Hewlett-Packard, USA
Timos Sellis   Institute for the Management of Information Systems
                               and National Technical University of Athens, Greece

Program Committee

Martin Bichler   Technical University of Munich, Germany
Christof Bornhoevd   SAP Labs, USA
Mike Carey   BEA, USA
Fabio Casati   University of Trento, Italy
Surajit Chaudhuri   Microsoft, USA
Dimitrios Georgakopoulos   Telcordia Technologies, USA
Jayant Haritsa   IISc, India
Howard Ho   IBM Almaden, USA
Tan Kian-Lee   National University of Singapore, Singapore
Wolfgang Lehner   University of Dresden, Germany
Torben B. Pedersen   Aalborg University, Denmark
Krithi Ramamritham   IIT Bombay, India
Stefano Rizzi   University of Bologna, Italy
Donovan Schneider   Yahoo, USA
Alkis Simitsis   Stanford University, USA
Panos Vassiliadis   University of Ioannina, Greece
Andrew Witkowski   Oracle, USA

Publication Chair

Alkis Simitsis   Stanford University, USA

Table of Contents

Situational Business Intelligence

Alexander Löser, Fabian Hueske, and Volker Markl

TU Berlin
Database System and Information Management Group
Berlin, Germany
`firstname.lastname@tu-berlin.de`

Abstract. Traditional business intelligence has focused on creating dimensional models and data warehouses, where after a high modeling and creation cost structurally similar queries are processed on a regular basis. So called "ad-hoc" queries aggregate data from one or several dimensional models, but fail to incorporate other external information that is not considered in the pre-defined data model. We focus on a different kind of business intelligence, which spontaneously correlates data from a company's data warehouse with "external" information sources that may come from the corporate intranet, are acquired from some external vendor, or are derived from the internet. Such situational applications are usually short-lived programs created for a small group of users with a specific business need. We will showcase the state-of-the-art for situational applications as well as the impact of Web 2.0 for these applications. We will also present examples and research challenges that the information management research community needs to address in order to arrive at a platform for Situational Business Intelligence.

Keywords: Business intelligence over text, Ad-hoc analysis, Cloud Computing.

1 Introduction

The long tail phenomenon is often observed with the popularity of consumer goods, web pages or tags, used in Flickr " or "MySpace". Interestingly, current enterprise applications are characterized by a long tail distribution as well: A small number of business critical applications are maintained by the IT-department. Such applications typically require high availability, high scalability and are requested by a large number of users. Examples for such business critical systems are mostly systems that manage business transactions e.g., accounting, *customer relationship management (CRM), enterprise resource planning (ERP)* and simple *online analytical processing application.* Besides of these business critical applications a "long tail" of *situational applications* exists. These are created to solve a short term problem and are often developed ad-hoc and independent from the IT-department. However, the growing amount of unstructured text in the web, intranets or user-generated content in blogs or reviews [RT07] needs to be integrated with structured information from a local ware house in an ad-hoc application. Neither conventional search engines nor conventional BI tools address this problem, which lies at the intersection of their capabilities. However, situational business intelligence applications tackle this problem. They tap

M. Castellanos, U. Dayal, and T. Sellis (Eds.): BIRTE 2008, LNBIP 27, pp. 1–11, 2009.

into the wealth of unstructured information in order to determine new trends and give an enterprise a competitive advantage. Major building blocks of situational business intelligence applications are information extraction algorithms and frameworks like UIMA [FL04], which identify relevant concepts and relationships within and between documents.

In situational business intelligence, the value of information decreases over time. Hence, the time for *a-priori* building semantic indexes may prevent BI users from getting a fast answer. Therefore another building block will be cloud computing. This technology enables an information worker to analyze and extract information *ad-hoc and at query time* from large corpora, such as a sample of several 10.000s documents from the web returned by a search web engine. Cloud architectures strive to massively parallelize complex computations on a large cluster through a computational model motivated by functional programming. Its scale-out and adaptability are exactly the kind of features needed to parallelize and scale-out UIMA aggregate analysis engines. Computing clouds provide highly available storage and compute capacity through distribution and redundancy. Most importantly, cloud computing architectures promise to adapt to changing requirements with respect to compute- and storage capacity by dynamically provisioning new compute or data nodes.

Our vision: We will develop a novel, database-inspired approach to ad-hoc analyze, aggregate and query very large data collections on cloud architectures. Our project builds upon the following five principles:

- **Common Algebraic Core.** We provide a common core of Situational Business Intelligence Applications - an extensible, algebraic model for executing complex queries over unstructured and (semi-) structured data. That model integrates data and text analysis operators and allows the analyst to describe, plan, optimize, and execute the proposed information extraction and query processing tasks.
- **Unstructured Text is a First Class Citizen**. Local analysis operations on a single document [RRK+08], global analysis operations on a set of documents [BCS+07] and entity ranking approaches for uncertain, extracted data [KSI+08, CYC07] must seamlessly fit into the algebraic framework and must be distributed efficiently on computing clouds.
- **Elementary Operators on the Cloud.** The algebra translates Situational Business Intelligence queries into functional data flow programs that are executed in a distributed fashion on a computing cloud. As starting point we base on elemental functional operators *map* and *reduce* and will add elementary operators, such as a *merge*-operator for combining multiple inputs, a *split*-operator for partitioning data and a *tree*-operator for improving efficiency of the relational join operation.
- **Query Optimization.** In order to minimize resource consumption or response time, we will identify beneficial heuristics and rules for rewriting initial query plans. We also need to identify parameters that impact execution to derive the preconditions under which the application of certain rewrite rules on the query plan is beneficial. Also, we will need to study how these parameters can be obtained efficiently before a particular query or a particular operation starts to execute.

- **Query Refinement.** Analysts need query refinement methods while selecting data sources, extractors, measurements and dimensions in an iterative query refinement process. We will investigate algorithms which determine alternative queries that serve the pragmatics of the querying user. Such an algorithm must determine the distance between the original query and some generated candidate query with respect to query semantics and coverage. Based on these tests it determines the candidate query whose results fit best the expectations of the querying user.

The rest of the paper is structured as follows: In Section 2 we propose typical use cases for business intelligence applications and identify major requirements, Section 3 reviews the building blocks of the system for running Situational Business Intelligence applications and outlines major research challenges that need to be solved before building such a system.

2 Situational Business Intelligence

The next generation of business intelligence applications requires analyzing and combining large data sets of both structured and unstructured data. According to current estimates, professional content providers (webmasters, journalists, technical writers, etc) produce about 2GB of textual content per day, whereas all user-generated content on the web amounts to 8-10GB per day [RT07]. Especially textual data usually is not processed in any manner, i.e., is not cleansed or annotated with an analysis schema. In this section we introduce our example scenarios and to our query answering process.

2.1 Example Scenario

Companies more and more tap into the analysis of consumer responses of consumers on company driven forums or external blogging sites. Incentives for analyzing the user-generated content are to increase customer loyalty, to bring ideas into the company, to research the market or to amplify "word of mouth" marketing. Often such analysis activities are initiated by non-IT people, e.g., product or marketing managers. The following use case tackles the ad-hoc analysis of "Business to Customer" communities of an "ordinary" product manager:

Example 1: A reseller of electronic goods is analyzing the digital camera market during the last 12 months. Therefore the reseller poses the following query:

Select customer reactions (sentiments) for digital cameras featuring 7-9 megapixels and below 400 euro during the last 12 months. Use results from amazon.com, google.com, blogspot.com and dpreview.com as data source.

The system receives ca. 560.000 candidate pages from a web search engine and filters out pages which do not contain any sentiments and do belong to dpreview.com, blogspot.com or amazon.com. From the remaining pages, the system extracts camera attributes (megapixel, price etc.), sentiments (time, author, positive/negative) and relationships between them. To avoid sparse data sets in the sentiments, identical camera models and their attributes are merged. Repeated sentiments from the same user and for same camera are counted. Next, the cleansed sentiment data is merged with structured sales data records from an internal data ware house. Finally, the system presents the product manager a time shift diagram, where he could identify correlations between sentiments and sales during the last 12 months.

2.2   Answering SBI Queries

Answering Situational Business Intelligence queries requires a close interaction between components for gathering text data, for extracting structured data from text, for cleansing extracted data, for obtaining a schema from the extracted data and for processing the extracted data on top of the generated schema. In the following, we review each step in the answering process.

1.   *Ad-hoc Data Retrieval:* SBI queries access various diverse data sources such as the Internet, cooperative intranets, data warehouses or office documents. While databases allow the retrieval of information with simple SQL interfaces, web sources are not accessible in such a structured way. E.g., the given example requires access to a data warehouse holding sales data for cameras and needs to retrieve user comments from internet blogs. Common methods to access web content are crawlers which materialize web sites locally by following links, such as Nutch-Hadoop, or to use web query languages, such as YQL[1].

2.   *Ad-hoc Data Extraction:* Unstructured data like forum pages or blogs contain lots of valuable information. Information extraction techniques are required to transform this information into a semi-structured model and make it accessible. For our example query, camera makes and models as well as user opinions (sentiment analysis) need to be extracted from blogging websites.

3.   *Data Cleansing of Extracted Data:* Data provided by data extraction services is often of low quality. E.g., the chosen extractor might not capture the semantic of the blogger and therefore not all attributes of a complex entity could be identified. Often current extraction services such as "OpenCalais.com" do not fuse syntactical different entities (e.g., "Dell" and "Dell Inc.") to a common logical entity. Data cleansing are resolving unique entities [WNJ08] or filling up missing information e.g., from domain knowledge.

4.   *Schema Generation from Uncertain Extracted Data:* In contrast to a static schema in a data ware house, Situational Business Intelligence applications require a flexible view over the data that is no longer coupled neither on available data in the data warehouse nor to a common 'one size fits all' schema. When formulating such an ad-hoc query over unstructured data, the analyst needs to estimate data volume and quality, available dimensions, facts and measurements for the chosen unstructured data source at query definition time. Cleary, for most data sources, the analyst will not have enough information to incorporate such estimates into ones query and requires additional system support.

5.   *Query Processing:* The data integration step requires all data available in structured or semi-structured format. Traditional database operators like filter,

---

[1] Yahoo! Query Language: http://developer.yahoo.com/yql

join, grouping and aggregating operators are applied on that data. For our example, the extracted blog data is filtered for camera of a specific manufacturer. This information is grouped by model and positive and negative opinions are counted. Finally, this data is joined with sales data for cameras models from the data warehouse.

3 Building a SBI System

Based on the requirements for a SBI system in this section we show a draft of the architecture of a Situational Business Intelligence system. Later, we define research challenges we face when building such a system and present recent technologies which have the capabilities to overcome these problems.

3.1 Architecture and Components

A business intelligence system needs to analyze large amounts of data in an ad-hoc fashion. Large amount of data result from a web search or a web crawl, where in a worst case millions of web search results or hundreds of GB of user-generated content need to be analyzed. To worsen the situation a business intelligence system must be able to answer an ad-hoc query usually in a few seconds to a few minutes. To address these goals, the authors of [LHDB2008] provided a basic model and process for analyzing structured and unstructured user generated content in a business warehouse. We extended their findings towards an ad-hoc query processing model and defined an

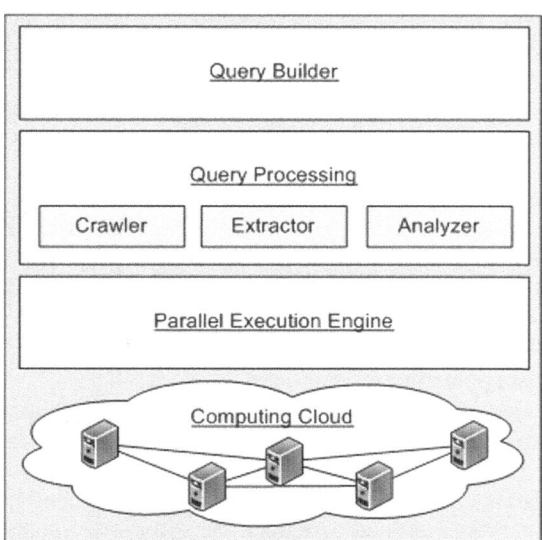

Fig. 1. Architecture of a situational business intelligence system

architecture model for a parallel data management system. In the following we list components for processing Situational Business Intelligence analytics queries.

Base Analytics: Enable Ad-hoc Queries but hide Complexity of Parallelization. Processing a large of data in a short amount of time is computationally intensive. The ability to parallelize, the fault-tolerance, and the adaptability of cloud architectures enable information management systems to answer queries on an internet-wide scale. However, the architecture of such a system is fundamentally different from a client/server architecture and storage subsystems, which traditional database systems utilize.

- **Cloud Computing.** Although Cloud Computing is not a fixed term, it describes network architectures with many lousily coupled computer nodes. These nodes are often built from commodity hardware. The property of loose coupling requires that cloud computing software must be able to cope with nodes joining or leaving the system, i.e. be fault tolerant and scalable. Due to the adaptability of the cloud, query optimization is mostly dynamic. Performance, cost, and energy consumption of the cloud guide adaptation of query execution plans at run-time.
- **Parallel Data Storage.** A distributed file system has to take the role of shared memory and storage system. Cached results and metadata have to be stored and indexed on the distributed file system of the cloud. Google's GFS [GGL03] or Microsoft's COSMOS [31] are optimized towards provide performance and fault tolerance. Hadoop[2], the open source MapReduce framework of the Apache Project, combines a distributed file system (HDFS) and a MapReduce runtime environment. However, current file systems do not provide random data access beyond file level. To overcome this limitation, column oriented data stores like Google's BigTable [CDG06] or HBASE have been built on top of distributed file systems. These stores organize data in a column oriented fashion which allows for simple extensions of storing schemata and sparse data, a feature which is crucial for ad-hoc text extraction tasks where often no schema is a-priori provided.
- **Parallel Execution Engine.** Due to the large number of nodes, Cloud Computing is especially applicable for parallel computing task working on elementary operations, such as *map* and *reduce*. MapReduce programs are executed in a runtime environment that is responsible for the distribution of computing tasks. These environments handle errors transparently and ensure the scalability of the system. [Dgh04] and [YDH07] provide details on the MapReduce programming paradigm and its applications.
- **Query Processor.** On cloud architectures, queries need to be translated into functional data flow programs. Above basic operations range from simple crawling over complex linguistic processing of a set of text documents to data analysis operations, such as value aggregation. Recently, data processing languages have been designed to reduce the efforts to implement data analysis applications on top of a distributed system. These languages are often tailored towards mass data analysis on flat data models, such as found at query log file

---

[2] http://hadoop.apache.org/

analysis. Examples for this type of query language are PigLatin [ORS08], Scope [CJL08] or the Cascading framework[3]. The language JAQL[4] supports a semi-structured data model and provides basic data analysis operations.

- **Query Builder.** Current query processing engines allow an analyst to model a data flow, which is executed as map-reduce program. The data flow is either specified as script language in a text document, such as with JAQL or PigLatin, or is included into a programming environment, such as in Cascading. However, current query builder do not provide specific operations for harvesting text data from the web, extracting entities and relationships and discovering a suitable analysis schema e.g., an OLAP-schema.

Advanced analytics: Run fast Analytics on Text Data. Particularly, complex ad-hoc queries over unstructured text require a data and processing model for a cloud architecture that seamlessly integrates advanced information extraction, and complex analysis operators with base analysis flow operators (filter, merge, join, aggregate, etc.). Functional map reduce programs can parallelize these advanced analysis operations but have to be integrated into the optimization framework of the overall system. We name requirements for executing advanced analysis operators on a map reduce platform:

- **Parallel Extraction leveraging Cost Models.** Information Extraction transforms unstructured textual information into semi-structured data and is a core component of a Situational Business Intelligence system. Current extraction systems [KKR06, SDN+07] combine local and global analysis techniques in an extraction plan. Parallelizing executing plans on a cloud platform would drastically reduce execution times. However, dependencies between "slower" and "faster" extractors during global analysis operations can cause bottlenecks. To avoid such bottlenecks each extractor provides a cost model. An optimizer will used it to distribute extraction load optimal over cloud nodes.
- **Parallel Testing of "Extractor - Data Source Combinations".** We except in the feature several providers of "text analytics" as service or by the domain independent information extraction paradigm [BCS+07]. A component is required that tests the quality and "fitness" of these services against text-data sources e.g., blogging web sites or news websites. To evaluate many potential "extractor–data source combination" fast, this component need to drastically parallelize the data gathering extraction and analysis process.
- **Distributed Data Cleansing.** Current text-analytics-as-service applications, such as OpenCalais.com, or UIMA annotators provide extracted but still "dirty" data snippets. We expect that new data cleansing techniques will benefit from additional context information in the text e.g., about the position of the extracted entities in the text. To leverage that information for large volumes of "user-generated-content", data cleansing techniques should be executed on a cloud.
- **Discovering Analysis Schemas from Uncertain, Extracted Data.** Schema discovery is the problem of constructing a relational schema that best describes extracted data. However, with text data, the analyst often is unfamiliar with the

---

[3] http://www.cascading.org/
[4] http://jaql.org/

data structure and cannot estimate the number of types and relationships a text will provide. Often the extracted data is inherently noisy, was extracted with a low precision or is ambiguous. Current research prototypes, such as NAGA [KSI+08], R-CUBE [PLAP08] or EntityRank [CYC07], use language models to rank in a top-k fashion entities and relationships from an extracted fact data base. Their probabilistic approaches consider uncertainty at the level of the extractor, at the level of the extracted span and the level of the document. A schema generation approach for Situational Business Intelligence applications needs to incorporate that uncertainty when generating the schema on a web-scale. "Starting points" are existing schema generation solutions for mining keys and foreign keys [SBHR06] for discovered high level structures in schemas [WRSM08] or for generating schemas from domain independent information extraction systems [CSE07].

3.2 Research Challenges

The challenge of executing Situational Business Intelligence analytics focuses on ad-hoc queries that neither traditional database management systems nor search engines could answer. In order to achieve that goal, we will address the following research challenges to improved base and advanced analytics:

Base Analytics: Extend language like JAQL, PIG or CASCADING as a single analysis-algebra for data gathering, extraction and processing. Unsolved research questions tackle the data and processing model in order to execute complex ad-hoc queries over structured and unstructured data. Can the model be formalized as a closed algebra? How can the model be translated efficiently to a cloud computing executing environment?

Base Analytics: Leveraging existing principles for distributed data management. Principles from Peer-to-Peer networks and Grid Computing techniques need to be transferred and improved for a cloud information management system. How does one perform query and storage load-balancing between the various processors in the cloud while not giving up data locality during query processing? How could a cloud run in a fault-tolerant way with high scalability and low cost? What management features are needed?

Advanced Analytics: Execute Relationship Extraction Techniques in Parallel. Common techniques to extract relationships base either on domain-dependent, rule-based systems or on domain-independent, open information extraction systems. Rule-based systems use dictionaries, grammars or extraction algebras to define the structure of entities and their relationships. However, the structure of entities and their relationships need to be known a-priori when formulating the rules for a specific text. Most rule-based tasks are local analysis tasks which could be executed as a map job and cloud be run in parallel execution mode. Open information extraction systems first train a classifier to detect relationships between entities on a set of documents. The classifier receives a set of entities (e.g., two company names for detecting an acquisition) and outputs a set of common grammatical and syntactic patterns (verbs, participles etc.) between these entities in the text. Based on these patterns relationships between entities are found, that are not known a-priori or where used in the training phase. The training step of these systems applies local analysis, such as detecting relationships in the training process per documents which could be executed as a map operation. However, training the classifier also includes and global analysis techniques to aggregate relationship patterns found on all documents within the corpus. Currently it is unclear on how to execute these global analysis operations and other problems e.g., web-wide key-generation from extracted data, data fusion of extracted data, anchor text analysis and home page search efficiently on a cloud. How can dependencies between extractors on global analysis tasks be optimized to guarantee a fast classifier building time and extraction time? How can we produce classifiers for millions of individual document collections on the fly?

Advanced analytics: On-demand Data Processing and Integration Operations. Typical advanced operations are time series or data cleansing operations. How can one flexibly extend the system with additional analysis operations on-demand? Often analyzed web data e.g., customer sentiments or data extracted from emails, needs to be "joined" with master data. How can one integrate operators for processing of structured and unstructured data sources into the cloud architecture?

Advanced analytics: End-user-driven Analysis. To drastically raise the number of users for analysts and lower the costs for developing the system infrastructure, ordinary "information worker" should have a simple access for posing ad-hoc analytics queries. Which new end-user driven analysis paradigms need to be offered? Can a dimensional model (in an OLAP sense) be derived ad-hoc from unstructured data? How can the user be guided to ask the right questions for deriving that model? How can ad-hoc user queries be reformulated based on available data and extraction operations? How can aggregation be performed and defined in a meaningful way, when the grouping criteria may not be known in advance, but are themselves extracted from the data?

Advanced Analytics: Data Flow Optimization across Analysis Operations. Currently, cloud computing languages, such as JAQL or Cascading do focus only on executing queries. Optimizations are conducted at the level of the MapReduce platform e.g., Hadoop. Given a system, where on-demand new operations are plugged in, how will the execution be parallelized, optimized and dynamically adapted? E.g., how does one combine or group crawling, extraction and analyzing data in order to efficiently process these operations on the same processor of the cloud? How do the operators have to be implemented to conduct optimizing of the entire flow? What is a "generic" cost model for such an operator?

4 Related Work

The internet is a source of lots of valuable information. There have been several attempts to leverage this data. The Cimple project [DSC07] provides a software platform which aims to integrate all information about specific user communities of the internet. The system starts with a high quality seed, consisting of initial data sources, relations and expert domain knowledge. From then on, it crawls its given web sources, extracts valuable information, integrates it and provides a web front end for browsing the joined information. Similar to Situational Business Intelligence systems, the Cimple platform autonomously crawls the web, extracts and integrates information. It differs from the concept of Situational Business Intelligence in that it is rather collecting information than allowing for extensive analyses. The Avatar project [KKR06] aims to provide semantic search over large data corpora such as corporate intranets or email archives. When answering a user query, the system is looking for documents whose content matches the intent of the search query. This requires three semantic analyses. First, the semantics of a document needs to be determined, second the intent of the query must be identified and finally the match between the documents semantic and the query intent must be checked. The matching step is done by a look-up in a semantic search index, which is build a-priori by applying information extraction techniques on all documents of the corpus. In contrast to Situational Business Intelligence systems the Avatar system does not analyze documents is an ad-hoc fashion. Furthermore, it only provides search functionality rather than complex analysis features. [BDJ07] presents methods to compute OLAP queries over uncertain data. Such data might result from applying information extraction techniques. Therefore, these methods might prove beneficial in the context of Situational Business Intelligence.

5 Conclusion

We have introduced a novel a class of applications for answering Situational Business Intelligence queries over web data. To answer such queries in an ad-hoc and fast fashion for samples including 10000s of web documents, we have introduced how cloud computing techniques need to be incorporated with text analytics, query processing and query refinement methods. We have named a few projects investigating how to integrate text analytics and query processing on top of extracted data. The next step is to drastically increase execution speed of these algorithms. The envisioned path for ad-hoc OLAP style query processing over textual web data may take a long time to mature. In any case, it is an exciting challenge that should appeal to and benefit from several research communities, most notably, the database, text analytics and distributed system worlds.

References

[BCS+07] Banko, M., Cafarella, M.J., Soderland, S., Broadhead, M., Etzioni, O.: Open Information Extraction from the Web. In: IJCAI 2007 (2007)
[BDJ07] Burdick, D., Deshpande, P.M., Jayram, T.S., Ramakrishnan, R., Vaithyanathan, S.: OLAP Over Uncertain and Imprecise Data. VLDB Journal 16(1) (January 2007)
[CJL08] Chaiken, R., Jenkins, B., Larson, P., Ramsey, B., Shakib, D., Weaver, S., Zhou, J.: SCOPE: Easy and Efficient Parallel Processing of Massive Data Sets. In: VLDB 2008 (2008)

[CRS08] Cooper, B., Ramakrishnan, R., Srivastava, U., Silberstein, A., Bonannon, P., Jacobsen, H., Puz, N., Weaver, D., Yerneni, R.: PNUTS: Yahoo!'s Hosted Data Serving Platform. In: VLDB 2008 (2008)

[CDG06] Chang, F., Dean, J., Ghemawat, S., Hsieh, W., Wallach, D., Burrows, M., Chandra, T., Fikes, A., Gruber, R.: Bigtable: A Distributed Storage System for Structured Data. In: OSDI 2006 (2006)

[CSE07] Cafarella, M., Suciu, D., Etzioni, O.: Navigating Extracted Data with Schema Discovery. In: WebDB 2007 (2007)

[CYC07] Cheng, T., Yan, X., Chen-Chuan Chang, K.: EntityRank: Searching Entities Directly and Holistically. In: VLDB 2007, pp. 387–398 (2007)

[DG04] Dean, J., Ghemawat, S.: Map Reduce: Simplified Data Processing on Large Clusters. In: OSDI 2004 (2004)

[DSC07] DeRose, P., Shen, W., Chen, F., Doan, A., Ramakrishnan, R.: Building Structured Web Community Portals: A Top-Down, Compositional, and Incremental Approach. In: VLDB 2007 (2007)

[FL04] Ferrucci, D., Lally, A.: UIMA: an architectural approach to unstructured information processing in the corporate research environment. Natural Language Engineering 10(3-4) (September 2004)

[Gart08] Gartner Executive Programs CIO Survey 2008 (January 10, 2008)

[GGL03] Ghemawat, S., Gobioff, H., Leung, S.: The Google file system. In: SOSP 2003 (2003)

[GS04] Götz, T., Suhre, O.: Design and implementation of the UIMA Common Analysis System. IBM Systems Journal 43(3) (2004)

[IBY+07] Isard, M., Budiu, M., Yu, Y., Birrell, A., Fetterly, D.: Dryad: Distributed Data-Parallel Programs from Sequential Building Blocks. In: EuroSys 2007 (2007)

[KKR06] Kandogan, E., Krishnamurthy, R., Raghavan, S., Vaithyanathan, S., Zhu, H.: Avatar semantic search: a database approach to information retrieval. In: SIGMOD 2006 (2006)

[KSI+08] Kasneci, G., Suchanek, F.M., Ifrim, G., Ramanath, M., Weikum, G.: NAGA: Searching and Ranking Knowledge. In: ICDE 2008 (2008)

[ORS08] Olston, C., Reed, B., Srivastava, U., Kumar, R., Tomkins, A.: Pig Latin. A Not-So-Foreign Language for Data Processing. In: Sigmod 2008 (2008)

[PLAP08] Pérez, J.M., Llavori, R.B., Aramburu, M.J., Pedersen, T.B.: Integrating Data Warehouses with Web Data: A Survey. IEEE Trans. Knowl. Data Eng. 20(7), 940–955 (2008)

[RRK+08] Reiss, F., Raghavan, S., Krishnamurthy, R., Zhu, H., Vaithyanathan, S.: An Algebraic Approach to Rule-Based Information Extraction. In: ICDE 2008 (2008)

[RT07] Ramakrishnan, R., Tomkins, A.: Towards a PeopleWeb. IEEE Computer 40(8) (2007)

[SBHR06] Sismanis, Y., Brown, P., Haas, P.J., Reinwald, B.: GORDIAN: Efficient and Scalable Discovery of Composite Keys. In: VLDB 2006 (2006)

[SDN+07] Shen, W., Doan, A., Naughton, J.F., Ramakrishnan, R.: Declarative information extraction using datalog with embedded extraction predicates. In: VLDB 2007 (2007)

[WNJ08] Weis, M., Naumann, F., Jehle, U., Lufter, J., Schuster, H.: Industry-Scale Duplicate Detection. In: VLDB 2008 (2008)

[WRSM08] Wu, W., Reinwald, B., Sismanis, Y., Manjrekar, R.: Discovering topical structures of databases. In: SIGMOD Conference 2008 (2008)

[YDH07] Yang, H., Dasdan, A., Hsiao, R., Parker, D.: Map-Reduce-Merge: Simplified Relational Data Processing on Large Clusters. In: Sigmod 2007 (2007)

On Solving Efficiently the View Selection Problem under Bag-Semantics*

Foto Afrati[1], Matthew Damigos[1], and Manolis Gergatsoulis[2]

[1] Department of Electrical and Computing Engineering,
National Technical University of Athens (NTUA), 15773 Athens, Greece
{afrati,mgdamig}@softlab.ntua.gr
[2] Department of Archive and Library Sciences, Ionian University,
Ioannou Theotoki 72, 49100 Corfu, Greece
manolis@ionio.gr

Abstract. In this paper, we investigate the problem of view selection for workloads of conjunctive queries under bag semantics. In particular we aim to limit the search space of candidate viewsets. In that respect we start delineating the boundary between query workloads for which certain restricted search spaces suffice. They suffice in the sense that they do not compromise optimality in that they contain at least one of the optimal solutions. We start with the general case, where we give a tight condition that candidate views can satisfy and still the search space (thus limited) does contain at least one optimal solution. Preliminary experiments show that this reduces the size of the search space significantly. Then we study special cases. We show that for chain query workloads, taking only chain views may miss all optimum solutions, whereas, if we further limit the queries to be path queries (i.e., chain queries over a single binary relation), then path views suffice. This last result shows that in the case of path queries, taking query subexpressions suffice.

1 Introduction

The *view selection problem* has received significant attention in many data-management scenarios, such as information integration, data warehousing, website designs, and query optimization. The static version of this problem is to choose a set of views to materialize over a database schema, such that (a) the cost of evaluating a set of queries is minimized, and (b) the views fit into a prespecified storage space. In query optimization, evaluating a set of queries using previously materialized views can significantly speed-up query processing, as part of the computation necessary for each query may have been done while computing views. Moreover, a set of similar queries (e.g. queries with similar

* This paper is part of the 03EΔ176 research project, implemented within the framework of the "Reinforcement Programme of Human Research Manpower" (PENED) and co-financed by National and Community Funds (25% from the Greek Ministry of Development-General Secretariat of Research and Technology and 75% from E.U.-European Social Fund).

M. Castellanos, U. Dayal, and T. Sellis (Eds.): BIRTE 2008, LNBIP 27, pp. 12–28, 2009.

subexpressions) can be computed efficiently by selecting an appropriate set of views that exploits these sharing opportunities. In a data warehouse, a successful selection of views to materialize can preclude costly access to the base relations and consequently helps to answer a batch of queries in efficient way. Similarly, the choice of a proper set of views to precompute may improve the performance of web-sites; because the set of expected queries can be answered quickly [10].

In contrast to the query answering problem using views, where the set of views is initially given, the view selection problem indicates automated techniques to produce the appropriate set of materialized views. In this paper, we focus on the view selection problem using query rewriting techniques, assuming that both query and view definitions are conjunctive queries. We use bag-semantics, which means that duplicate occurrences of tuples are allowed to query answers and to database relations [19]. The "bag-approach" of the problem is more practical because of its close relationship to the SQL features where bag-relations are allowed and the duplicate tuples are not eliminated during the query evaluation; unless explicitly requested (by using the DISTINCT keyword).

The hardness of the view selection, as defined and investigated in [8,2,7,12], is caused by the bicriteria nature of the problem. These criteria are: (1) for a given set of views, the selection of the less-costly equivalent rewritings of the queries and (2) the choice of the appropriate set of views which does not violate the storage constraint. Bicriteria settings have different variants (and consequently different solutions and complexity results) depending of which of the two objective functions of these two criteria is required to be optimized under the constraint that the value of the other objective function does not exceed a bound that is given by the designer of the system. In this paper, we consider the variant of the problem in which we want to find a viewset such that it does not exceed the storage constraint and is optimum with respect to the evaluation cost of the query workload. We count the size of a viewset as the number of tuples required to store all the views in the viewset (however we notice that all our results hold under a more general count) and the cost of query evaluation is based on the sum-of-joins cost model for left-linear query plans (the exact definitions can be found in subsequent section). In [2] the same problem is investigated and is shown that we can restrict the search space for views in the viewset only to those views that are *generalizations of query subexpressions*.

Our contributions in this paper are: a) In Section 4.1, we improve the search space of [2] by showing that it suffices to consider only the *least general generalizations of query subexpressions*. In particular, we show that if we restrict ourselves to this smaller search space, an optimal solution is always retained, i.e., a solution which satisfies the storage limit and achieves the optimum value for the evaluation cost. b) Based on these results, we develop in Section 4.2 an efficient algorithm for finding an optimal solution. c) We study (Section 5) the problem for two special cases, namely when all queries in the workload are chain queries and when they are path queries. In the first case we show that we cannot restrict the search space to only chain views because we may loose all optimal solutions (Section 5.1). In the second case, we show that we can restrict further the search space to consider only path views (Section 5.2).

2   Related Work

The problem of automatic selection of views to materialize has attracted the interest of many researchers. In [7], the space requirements for the view selection problem in the context of data warehouse design under set-semantics, are considered. This paper, also investigates conditions under which the search space of optimal configurations can be reduced to the views that are subexpressions of the queries in the workload. In [21,23], the extraction of common subexpressions of the queries in the workload is studied. The authors in [21], study the problem of searching for a maximum common subexpression of a workload, while [23] proposes an algorithm for searching for maximum common subexpressions for a subclass of select-project-join SQL queries, using query graphs. Another approach for finding similar subexpressions is proposed in [25] where workloads of select-project-join-groupby queries are considered. The authors propose a solution for the multi-query optimization problem which is incorporated in the Microsoft SQL Server. The algorithm has a lightweight mechanism (table-signatures) to detect common subexpressions and multiple sharing opportunities.

In [12], it is stated the view selection problem using AND-OR graphs to represent the query plans. Two types of constraints on materialized views are assumed, a storage limit and a maintenance-cost constraint. The candidate set of view configurations are given as input, hence the time of the construction of view configurations is not considered in the response time of the algorithms.

In [20], the view selection problem assuming a maintenance-cost constraint in the data warehouse environment and proposed an algorithm based on multi-query graphs, is studied. In [24], the authors examine greedy/heuristic algorithms for solving the view-selection problem assuming a maintenance-cost constraint and OLAP queries in multidimensional data warehouse environment. In [6] the problem for multidimensional databases is studied and an algorithm that selects views by reducing significantly the solution space is proposed; considering only the relevant elements of the multidimensional lattice. The authors considered the standard SQL notion of group-by and aggregate functions in order to capture queries with aggregation. In earlier work, Rizzi and Saltarelli [18] presented a comparative evaluation that uses view materialization and indexing for a single GSPJ (Group-by-Select-Project-Join) query expressed on a star scheme for the data warehousing context.

The view selection problem, in the context of multidimensional data warehouses, also studied by several authors [14,13,11]. In [14], it is described a system which was incorporated in Microsoft SQL Server and focuses on selection of both views and indexes. Earlier, the authors of [13] propose algorithms for selecting views in the case of data cubes and study the complexity of the problem. In [11], the work of [13] was further extended to include index selection.

A significant result that underlines the difference of the view selection problem in the case of queries with and without aggregation is presented in [3]. In this work, an algorithm for selecting views is proposed and complexity results are presented, using a theoretical approach to express GSPJ queries. The authors also showed that using materialized views to compute aggregate queries results greater benefits than for purely conjunctive queries; as a view with aggregation precomputes some of the grouping/aggregation on some of the query's subgoals.

In [8], Chirkova et al. observed that the complexity of view selection problem under set semantics, and assuming conjunctive query workload, depends crucially on the quality of the estimates that a query optimizer has on the size of views. In [8], it is also shown that an optimal choice of views may involve an exponential number of views in the size of the database schema. In the same context, in [2], Afrati et al. study the search space of candidate sets of views, under bag, set and bag-set semantics. Finally, the problem of selecting minimal-size-views to materialize has been studied theoretically in [9], where the problem has been proven decidable and an upper bound is given on this problem's complexity.

3   Preliminaries

3.1   Basic Definitions

A *relation schema* is a named relation defined by its name R (called *relation name*) and a set A of *attributes*. A *relation instance* r for a relation schema is a collection of tuples over its attribute set. The schemas of the relations in a database constitute its *database schema*. A *relational database instance* (*database*, for short) is a collection of stored relation instances. A relation instance can be viewed either as a *set* or as a *bag* (or *multiset*) of tuples. A *bag* (or bag-relation [22]) can be thought of as a set of elements with multiplicities attached to each element. In a *set-valued database*, all stored relations are sets; in a *bag-valued database*, multiset stored relations are allowed. The *bag-operators* [22] are similar to the set-operators. The difference is that in bag-selection and bag-projection duplicate tuples in the result are not eliminated. Concerning the Cartesian product, the difference is that the multiplicity of each tuple t obtained in $R \times S$ from a tuple t_1 of R and a tuple t_2 of S is $m \cdot n$, where m is the multiplicity of t_1 and n is the multiplicity of t_2. Depending on whether a database is bag or set-valued and the operators are set or bag operators, the queries may be computed under *set-semantics* (considering set-valued databases and operators), *bag-semantics* (considering bag-valued databases and operators), or *bag-set semantics* (considering set-valued databases and bag-operators). We consider bag-semantics in this paper.

A *query* is a mapping from databases to databases, usually specified by a logical formula on the schema $\mathcal{S}$ of the input databases. Typically, the output database (called *query answer*) is a database with a single relation. In this paper we focus on the class of select-project-join SQL queries with equality comparisons, a.k.a. safe *conjunctive queries* (CQs for short). Formally, a conjunctive query definition [1] is a rule of the form:

$$Q : q(\overline{X}) :\text{-} g_1(\overline{X}_1), \ldots, g_n(\overline{X}_n)$$

where $g_1, \ldots, g_n$ are database relations and $\overline{X}, \overline{X}_1, \ldots, \overline{X}_n$ are vectors of variables or constants. The atom $q(\overline{X})$ is the *head* of Q while the atoms on the right of :- are said to be the *body* of Q. Each $g_i(\overline{X}_i)$ is also called a *subgoal* of Q. The variables in $\overline{X}$ are called *distinguished* or *head variables* of Q, whereas the variables in $\overline{X}_i$ are called *body variables* of Q. A body variable which is not also a head variable is called *non-distinguished variable* of Q. In this work, we consider *safe conjunctive queries* that is CQs whose head variables also occur in their body. A *chain query* is a conjunctive query of the following form:

$$Q : q(X_0, X_n) :\text{-} r_1(X_0, X_1), r_2(X_1, X_2), \ldots, r_n(X_{n-1}, X_n)$$

where $r_1, \ldots, r_n$, are binary relations and $X_0, X_1, \ldots, X_n$ are variables. If the relation symbols $r_1, \ldots, r_n$ are identical then the query is called *path query* of length n, denoted as P_n. A *view* refers to a named query. A view is said to be *materialized* if its answer is stored in the database. In this work, we are restricted to the use of views defined by conjunctive queries called *conjunctive views*.

A *substitution* θ [15] is a finite set of the form $\{X_1/Y_1, \ldots, X_n/Y_n\}$, where each Y_i is a variable or a constant, and $X_1, \ldots, X_n$ are distinct variables. When $Y_1, \ldots, Y_n$ are distinct variables, θ is called *renaming substitution*. In the following we also use the notion of *expression* to denote a conjunction of atoms. Let $\theta = \{X_1/Y_1, \ldots, X_n/Y_n\}$ be a substitution. Then the *instance $E\theta$* of an expression (resp. a query) E, is the expression (resp. the query) obtained by simultaneously replacing each occurrence of X_i in E by Y_i for all $i = 1, \ldots, n$.

Definition 1. *An expression E is a generalization of an expression E' if E' is an instance of E. E is a common generalization of $E_1, \ldots, E_n$, with $n > 1$ if E is a generalization of each expression E_i, with $1 \leq i \leq n$. E is a least common generalization (or a least general generalization - lgg [16]) of $E_1, \ldots, E_n$, with $n > 1$, if E is a common generalization of $E_1, \ldots, E_n$, and for each common generalization G of $E_1, \ldots, E_n$, the expression G is a generalization of E.*

3.2   Query Rewriting and the View Selection Problem

Given a set of views (also, called *viewset*) $\mathcal{V}$ defined on a database schema $\mathcal{S}$, and a database $\mathcal{D}$ on the schema $\mathcal{S}$, then by $\mathcal{V}(\mathcal{D})$ we denote the database obtained by computing all the view relations in $\mathcal{V}$ on $\mathcal{D}$. Moreover, let Q be a query defined on $\mathcal{S}$. A query R is a *rewriting* of the query Q using the views in $\mathcal{V}$ if all subgoals of R are view atoms defined in $\mathcal{V}$. The *expansion R^{exp}* of R is obtained by replacing all view atoms in the body of R with their corresponding base relations. Non-distinguished variables in a view definition are replaced with fresh variables in R^{exp}. A rewriting R of a query Q on a viewset $\mathcal{V}$ is an *equivalent rewriting* if $R(\mathcal{V}(\mathcal{D})) = Q(\mathcal{D})$, for every database $\mathcal{D}$. In [19], it is proved that a rewriting R of a query Q, under bag-semantics, is equivalent to Q if and only if there is an one-to-one containment mapping from Q to the R^{exp}.

Given a set $\mathcal{Q}$ of queries (also called *query workload*), defined on a schema $\mathcal{S}$, and a database instance $\mathcal{D}$, we want to find and precompute offline a viewset

$\mathcal{V}$ defined on $\mathcal{S}$, such that the views in $\mathcal{V}$ can be used to compute the answers to all queries in the workload $\mathcal{Q}$. More specifically, our problem, called the *view selection problem*, is to find a set of views that when materialized, (a) would satisfy a set $\mathcal{L}$ of constraints on the size of the views, and (b) can be used to get equivalent rewritings of the queries in $\mathcal{Q}$ which minimizes the evaluation cost of the queries. We refer to the tuple $\mathcal{P} = (\mathcal{S}, \mathcal{Q}, \mathcal{D}, \mathcal{L})$ as the *input of view selection problem*. The view selection problem is said to be *bag-oriented* (resp. *set-oriented*, or *bag-set-oriented*) if we consider bag semantics (resp. set semantics, or bag-set semantics).

In this paper, we consider that the only constraint on materialized views is a storage limit L (i.e. $\mathcal{L} = \{L\}$), which is a bound on the size of the views (which represents the available disk space for storing the views). Our goal is to choose the viewsets which minimize the evaluation cost of the queries and whose size will not exceed the limit L. Notice that, if the storage limit is sufficiently large then we can materialize all query answers, which is an optimal viewset. The problem becomes interesting when the storage limit is less than that. In the following we measure the size of a relation R as the number of tuples in R.

Definition 2. *Let $\mathcal{P} = (\mathcal{S}, \mathcal{Q}, \mathcal{D}, \mathcal{L})$ be a view selection problem input. A viewset $\mathcal{V}$ is said to be* admissible *for $\mathcal{P}$ if (1) $\mathcal{V}$ gives equivalent (candidate) rewritings of all the queries in $\mathcal{Q}$, (2) for every view $V \in \mathcal{V}$, there exists at least one equivalent rewriting of a query in $\mathcal{Q}$ that uses V, and (3) $\mathcal{V}$ satisfies the constraints $\mathcal{L}$.*

The following definition formally defines the *solution* and *optimal solution* of view selection problem for a given input.

Definition 3. *Let a view selection problem input $\mathcal{P} = (\mathcal{S}, \mathcal{Q}, \mathcal{D}, \mathcal{L})$.*

- *A* solution *of $\mathcal{P}$ is a tuple $(\mathcal{V}_{adm}, \mathcal{R})$, where $\mathcal{V}_{adm}$ is an admissible viewset for $\mathcal{P}$ and $\mathcal{R}$ is a set of equivalent rewritings of the queries in $\mathcal{Q}$ using $\mathcal{V}_{adm}$.*
- *An* optimal solution *for $\mathcal{P}$ is a solution which minimizes the cost of evaluating the queries in the workload among all solutions of $\mathcal{P}$. The viewset in an optimal solution is said to be an* optimal viewset.

Optimal solutions relate to the estimation of the cost of evaluating a query. We thus demand from the optimal solutions to minimize a given cost-function that we employ. We assume that the view relations have been precomputed, hence we do not assume any cost of computing the views. For conjunctive queries we use the *sum-of-joins* cost model which measures the cost of query evaluation as the sum of the costs of all the joins in the evaluation. More precisely, suppose we are given a query Q and a database $\mathcal{D}$. We assume use of only *left-linear query plans*, where selections are pushed as far as they go and projection is the last operation. Thus, each plan is a permutation of the subgoals of the query, and the cost of this query plan on a given database instance $\mathcal{D}$ is defined inductively as follows. For $n = 1$, the cost of query plan $Q = R_1$ is the size of the relation R_1. For each $n \geq 2$, the cost of query plan $(\dots((R_1 \bowtie R_2) \bowtie R_3) \bowtie \dots \bowtie R_n)$ over n relations is the sum of the following four values:

1. the cost of query plan $(\ldots((R_1 \bowtie R_2) \bowtie R_3) \bowtie \ldots \bowtie R_{n-1})$
2. the size of relation $R_1 \bowtie \ldots \bowtie R_{n-1}$
3. the size of relation R_n and
4. the size of relation $R_1 \bowtie \ldots \bowtie R_n$

The cost of evaluating a query Q on a database $\mathcal{D}$, denoted as $C(Q, \mathcal{D})$, is the minimum cost over all Q's query plans when evaluated on $\mathcal{D}$. Moreover, the cost of a query workload, denoted as $C(\mathcal{Q}, \mathcal{D})$, is defined as the sum of the costs of all queries in the workload. In this paper, although we use the above cost model, our results also hold for cost-models for which the evaluation cost is increasing with the size of intermediate relations [11,8,2,4].

4   The Space of Optimal Solutions

In this section, we elaborate on the search space analysis of candidate solutions for bag-oriented view selection problems, considering that both queries and views are conjunctive queries/views. The main results of this section are as follows: In Subsection 4.1, we propose techniques to reduce the search space of candidate views and demonstrate that if there exists a solution for a given problem input, then there is at least one optimal solution of a specific form. We refer to these solutions as the *representative (optimal) set of solutions*. In Subsection 4.2, an algorithm is presented that computes the representative set of optimal solutions.

4.1   Representative Set of Solutions

In [2], it has been proved that for workloads of conjunctive queries each view in any admissible viewset (and thus in any optimal viewset) can be defined as a generalization of a subexpression of some query in the workload. The following lemma, which combines Lemmas 2 and 3 of [2], presents this result formally:

Lemma 1. *Let $\mathcal{P} = (\mathcal{S}, \mathcal{Q}, \mathcal{D}, \mathcal{L})$ be a conjunctive bag-oriented problem input, $\mathcal{V}$ be any* admissible *viewset for $\mathcal{P}$, and Q be any query in $\mathcal{Q}$. Suppose that $\mathcal{V}' \subseteq \mathcal{V}$ is the set of all views used in an equivalent rewriting R of Q in terms of $\mathcal{V}$. Then:*

1. *The subgoals in the expansion of R corresponding to the definitions of views $\mathcal{V}'$ form a partition of the (subgoals in the) definition of Q.*
2. *Each view in $\mathcal{V}'$ can be defined as a generalization of a subexpression of Q which is a member of the partition as defined in (1).*

Lemma 1 precisely describes a search space (consisting of all query subexpressions and their generalizations) to look for view definitions. As, in general, this search space is huge, it is crucial to investigate ways to reduce this search space (possibly for special cases of the view selection problem) in order to construct efficient algorithms for solving the view selection problem. A significant improvement in this direction might be to restrict the search space to contain only the subexpressions of the queries in the query workload (i.e. to exclude the generalizations of the subexpressions). Unfortunately, as it is shown in the following example, in the general case this is not possible.

Example 1. Consider a database schema $\mathcal{S}$ that contains only the relation e of arity 4 and a query workload $\mathcal{Q} = \{Q_1, Q_2\}$ on $\mathcal{S}$, where:

$$Q_1 : \ q_1(X,Y) :\text{-} \ e(X,X,X,Y).$$
$$Q_2 : \ q_2(X,Y) :\text{-} \ e(X,Y,Y,Y).$$

Consider also the following three viewsets $\mathcal{V}_1$, $\mathcal{V}_2$ and $\mathcal{V}_3$:

- $\mathcal{V}_1 = \{V_{11}, V_{12}\}$, where:
$$V_{11} : v_{11}(X_1, X_2) :\text{-} \ e(X_1, X_1, X_1, X_2).$$
$$V_{12} : v_{12}(X_1, X_2) :\text{-} \ e(X_1, X_2, X_2, X_2).$$
- $\mathcal{V}_2 = \{V_2\}$, where:
$$V_2 : v_2(X_1, X_2, X_3) :\text{-} \ e(X_1, X_2, X_2, X_3).$$
- $\mathcal{V}_3 = \{V_3\}$, where:
$$V_3 : v_3(X_1, X_2, X_3, X_4) :\text{-} \ e(X_1, X_2, X_3, X_4).$$

Notice that the bodies of the view definitions of $\mathcal{V}_1$ are subexpressions of the bodies of the queries in $\mathcal{Q}$ (in fact they are obtained from the bodies of Q_1 and Q_2 by renaming their variables), while the bodies of the views in $\mathcal{V}_2$ and $\mathcal{V}_3$ are generalizations of these subexpressions. Using each one of the above viewsets we get equivalent rewritings for the queries in $\mathcal{Q}$. More specifically, using $\mathcal{V}_1$ we get:

$$R_1 : r_1(X,Y) :\text{-} \ v_{11}(X,Y).$$
$$R_2 : r_2(X,Y) :\text{-} \ v_{12}(X,Y).$$

where R_1 and R_2 are equivalent rewritings of Q_1 and Q_2 respectively. Using $\mathcal{V}_2$ we get:

$$R_1' : r_1'(X,Y) :\text{-} \ v_2(X,X,Y).$$
$$R_2' : r_2'(X,Y) :\text{-} \ v_2(X,Y,Y).$$

where R_1' and R_2' are equivalent rewritings of Q_1 and Q_2 respectively. Finally, using $\mathcal{V}_3$ we get:

$$R_1'' : r_1''(X,Y) :\text{-} \ v_3(X,X,X,Y).$$
$$R_2'' : r_2''(X,Y) :\text{-} \ v_3(X,Y,Y,Y).$$

where R_1'' and R_2'' are equivalent rewritings of Q_1 and Q_2 respectively.

Assuming a database instance D=\{(e(a, a, a, a);1), (e(a, b, c, d);5)\}, the sets $\mathcal{V}_1(D)$, $\mathcal{V}_2(D)$ and $\mathcal{V}_3(D)$ are:

$$\mathcal{V}_1(D) = \{(v_{11}(a,a); 1), (v_{12}(a,a); 1)\}.$$
$$\mathcal{V}_2(D) = \{(v_2(a,a,a); 1)\}.$$
$$\mathcal{V}_3(D) = \{(v_3(a,a,a,a); 1), (v_3(a,b,c,d); 5)\}.$$

Since $size(\mathcal{V}_3(D))=6$, $size(\mathcal{V}_1(D))=2$ and $size(\mathcal{V}_2(D))=1$, we have $size(\mathcal{V}_3(D)) > size(\mathcal{V}_1(D)) > size(\mathcal{V}_2(D))$. If we choose a storage limit $L = size(\mathcal{V}_2(D)) = 1$, then $\mathcal{V}_2$ is the only admissible viewset among the above three.

Example 1 shows that, in some cases, any optimal solution requires views that cannot be constructed as subexpressions of the queries in the query workload.

The optimal solution in Example 1 uses views constructed using generalizations of subexpressions of the queries. In particular, the view in the optimal viewset $\mathcal{V}_2$ is defined as a common generalization of the bodies of both queries in the query workload $\mathcal{Q}$. Based on these observations two questions arise:

1. Are there any special cases of the view selection problem for which there are optimal solutions whose viewset can be constructed by considering only subexpressions of the queries in the query workload?
2. For the general case, can we reduce the search space specified by Lemma 1 which consists of all possible generalizations of query subexpressions?

Both questions can be answered affirmatively as shown in the following Propositions 1 and 2.

Proposition 1. *Let $\mathcal{P} = (\mathcal{S}, \mathcal{Q}, \mathcal{D}, \mathcal{L})$ be a conjunctive bag-oriented view selection problem input such that every relation in $\mathcal{S}$ appears at most once in a body of some query in $\mathcal{Q}$. If there exists a solution for $\mathcal{P}$, then there exists an optimal solution $\Lambda = (\mathcal{V}, \mathcal{R})$ such that each view in $\mathcal{V}$ is defined as a subexpression of a query in $\mathcal{Q}$.*

Notice that, when the assumptions of Proposition 1 hold, the queries in the workload $\mathcal{Q}$ do not contain self-joins. In this case, because of Theorem 5 of [2], we can rewrite each query in $\mathcal{Q}$ without using self-joins of views in $\mathcal{V}$.

We now focus on the general case and prove, in Proposition 2, that, in order to construct an optimal viewset, we need to consider both subexpressions of queries and lgg's of subexpressions. We can thus exclude all those generalizations of subexpressions that are not lgg's of two or more subexpressions.

Proposition 2. *Let $\mathcal{P} = (\mathcal{S}, \mathcal{Q}, \mathcal{D}, \mathcal{L})$ be a conjunctive bag-oriented view selection problem. If there exists a solution for $\mathcal{P}$, then there is an optimal solution $\Lambda = (\mathcal{V}, \mathcal{R})$ for $\mathcal{P}$ such that the body of each view in $\mathcal{V}$ is either a subexpression of a query in $\mathcal{Q}$ or an lgg of two or more subexpressions of queries in $\mathcal{Q}$.*

The intuition behind Propositions 1 and 2 is that the use of generalization of subexpressions in defining a view is useful only when this view definition will be subsequently used two or more times to construct equivalent rewritings for the queries in the workload $\mathcal{Q}$. This is the case of the viewsets $\mathcal{V}_2$ and $\mathcal{V}_3$ in Example 1. Besides, it is not useful to generalize the subexpression more than needed as this, in general, increases the number of the tuples obtained when materializing this "overgeneralized" view definition and this does not contribute towards an improvement of the evaluation of the rewriting . An example of such "overgeneralization" is the viewset $\mathcal{V}_3$ in Example 1.

We further refine Propositions 1 and 2 by restricting also the vector of variables in the heads of the view definitions. The simplest choice is to put as arguments of a view head all different variables appearing in the view's body. However, this is not always the "best" choice as the following example shows:

Example 2. Consider a query workload $\mathcal{Q} = \{Q\}$, where:

$$Q: \quad q_1(X, Y) :\text{-} e(X, Z), f(Z, W), g(W, Y).$$

Consider also the following viewset $\mathcal{V}_1 = \{V_{11}, V_{12}\}$:

$$V_{11} : v_{11}(X, Z, W) :\text{-} e(X, Z), f(Z, W).$$
$$V_{12} : v_{12}(W, Y) :\text{-} g(W, Y).$$

Notice that using $\mathcal{V}_1$ as we can get the following equivalent rewriting R of Q:

$$R : r(X, Y) :\text{-} v_{11}(X, Z, W), v_{12}(W, Y).$$

It is easy to see, however, that the variable Z in the head of V_{11} is redundant. More specifically, if we replace the view V_{11} in $\mathcal{V}_1$ by the following view V'_{11}:

$$V'_{11} : v'_{11}(X, W) :\text{-} e(X, Z), f(Z, W).$$

we get R' which is also an equivalent rewriting of Q:

$$R' : r'(X, Y) :\text{-} v'_{11}(X, W), v_{12}(W, Y).$$

Comparing V_{11} and V'_{11}, it is easy to see that, under bag semantics for every database D we have $size(\mathcal{V}_{11}(D)) = size(\mathcal{V}'_{11}(D))$. Also, the query R', obtained by using V'_{11} to rewrite Q, is computed more efficiently than the rewriting R obtained by using V_{11} to rewrite Q.

We now show how to choose the appropriate set of variables to be used as head arguments of the view definitions.

Definition 4. *Let Q be a query of the form $H :\text{-} B_1, \ldots, B_n$ and $S = B_1, \ldots, B_{1k}$, with $1 \le k \le n$, be a subexpression of the body of Q. Let $Q' = Q - S$ be the query obtained by removing from the body of Q the atoms in S. Then, the set $lvars(Q, S) = Vars(Q') \cap Vars(S)$, is called the* linking variables *of Q and S.*

Example 3. (Continued from Example 2) Consider the query Q in Example 2 and the subexpression $S = e(X, Z), f(Z, W)$ of Q. It is easy to see that the set of linking variables of Q and S is $lvars(Q, S) = \{X, W\}$.

Proposition 3. *Let Q be a conjunctive query and V be a view whose body is defined as a subexpression of Q. Then the view V can be used in an equivalent rewriting of Q, if and only if $lvars(Q, S) \subseteq vars(head(V))$.*

The linking variables are related to the shared-variables property introduced by [17]; that holds in the set-oriented context.

What the above proposition indicates is that the set of linking variables is the minimum set of variables that should be put in the head of the view definition so as this view can be used in an equivalent rewriting of the query.

Example 4. (Continued from Example 3) Notice that the variables in $\{X, W\}$, which are the linking variables of Q and S, appear in the heads orf both views V_{11} and V'_{11} constructed from the subexpression S of Q. Observe that, if we remove X or W or both from the head of the view V_{11} (or the view V'_{11}), then the corresponding viewset cannot give equivalent rewriting for the query Q.

Proposition 3 refers to views which are defined as subexpressions of the queries in the query workload. We now investigate the problem of selecting the head arguments of the views defined as least general generalizations of subexpressions of queries. For this we need the following definition:

Definition 5. *Let* $E = \{S_1, \ldots, S_k\}$, *with* $k > 1$, *be a set of expressions, and* G *be their least general generalization (supposing that such an lgg exists). Let* S *be an expression in* E *and* M *be a mapping for the arguments of* S *to the arguments of* G *such that each argument of* S *in a position* (i, j), *where* i *is the order of an atom in* S *and* j *is the order of the argument in the* i-*th atom of* S, *maps to the argument which is in the position* (i, j) *on* G. *Let* X *be a variable in* $vars(S)$. *Then the* corresponding variable set *of* X *in* G *is defined as* $\{Y | X$ *appears in a position* (i, j) *of* S *and* Y *is the variable in the position* (i, j) *of* $G\}$.

Proposition 4 specifies the minimum set of variables that should be put in the head of a view defined as the lgg of two or more subexpressions.

Proposition 4. *Let* $Q_1, \ldots, Q_k$, *with* $k > 1$, *be (not necessarily different) queries in a query workload* $\mathcal{Q}$, *and let* $S_1, \ldots, S_k$, *be expressions such that* S_i *is a subexpression of* Q_i *for* $1 < i \leq k$. *Suppose that the least general generalization of* $S_1, \ldots, S_k$ *exists and that* V *is a view whose body is the least general generalization of* $S_1, \ldots, S_k$. *Then the view* V *can be used in an equivalent rewriting of* Q_i, *for all* $i = 1, \ldots, k$, *if and only if* $\bigcup_{i=1}^{k} L_i \cup \bigcup_{i=1}^{k} M_i \subseteq vars(head(V))$, *where (a)* L_i *is the union of the corresponding variable sets of the variables in* $lvars(Q_i, S_i)$, *and (b)* M_i *is the union of the corresponding variable sets of the variables in* S_i *whose corresponding variable sets are not singletons.*

Another way to construct the view V whose body is the least general generalization of the subexpressions S_1 and S_2 of two queries Q_1 and Q_2 respectively proceeds in two steps as follows:

1. We construct the views V_1 and V_2 using the subexpressions S_1 and S_2 respectively as bodies and the linking variables with Q_1 and Q_2 as head variables.
2. By considering V_1 and V_2 as queries we construct V with body the lgg of the bodies of V_1 and V_2 and with head variables the minimum set of variables specified by Proposition 4. V is said to be an *lgview* of the views V_1 and V_2.

This procedure can be easily generalized for more than two subexpressions.

An interesting question referring to lgviews is the following: "Does the inequality $size(V) \leq size(V_1) + size(V_2)$ always hold for the lgview V of two views V_1 and V_2?". Notice that, if the answer is "yes" for any bag-oriented view selection problem input, then whenever an lgview exists, the original views can be discarded eliminating in this way the search space for finding viewsets. Unfortunately, the inequality does not always hold, as the following example shows.

Example 5. Let a viewset $\mathcal{V} = \{V_1, V_2\}$, where the definitions of the views are:

$$V_1 : v_1(X, Z) :\text{-} p_1(X, X), p_2(X, Z).$$
$$V_2 : v_2(X, Z) :\text{-} p_1(X, Z), p_2(Z, Z).$$

where p_1 and p_2 are binary relations on the database schema $\mathcal{S}$. Consider also another viewset $\mathcal{W} = \{W\}$ whose view W is defined as:

$$W : w(A, B, C) :\text{-} p_1(A, B), p_2(B, C).$$

Notice that W is the lgview of the views in $\mathcal{V}$. Assuming the database instance:

$$\mathcal{D} = \{p_1(1, 1), p_1(1, 2), p_1(3, 4), p_2(1, 1), p_2(1, 2), p_2(2, 2), p_2(2, 3), p_2(4, 5)\},$$

in which the multiplicity of each database tuple in this example is 1 and for this we omit it, and materializing the views over this database we get:

$$\mathcal{V}(\mathcal{D}) = \{v_1(1, 1), v_1(1, 2), v_2(1, 1), v_2(1, 2)\}.$$
$$\mathcal{W}(\mathcal{D}) = \{w(1, 1, 1), w(1, 1, 2), w(1, 2, 2), w(1, 2, 3), w(3, 4, 5)\}.$$

It is easy to see that size($\mathcal{V}(\mathcal{D})$) < size($\mathcal{W}(\mathcal{D})$).

The following theorem summarizes the results of this section:

Theorem 1. *Let a bag-oriented view selection input* $\mathcal{P} = (\mathcal{S}, \mathcal{Q}, \mathcal{D}, \mathcal{L})$. *If there is a solution for* $\mathcal{P}$, *then there exists an optimal solution* $\Lambda = (\mathcal{V}, \mathcal{R})$ *such that each view in* $\mathcal{V}$ *is either a subexpression view or an lgview whose body is constructed as specified by Proposition 2, and whose head is constructed using the minimal set of variables specified by Propositions 3 and 4, respectively.*

Thus the class of solutions constructed as above is a *representative set of solutions* for a given bag-oriented view selection problem input $\mathcal{P}$.

4.2   LGG-VSB Algorithm

An algorithm, called LGG-VSB, which is based on the results of the previous section, and outputs the representative set of optimal solutions, for a given view selection problem input, is proposed in this section. LGG-VSB incorporates the results of the Theorem 1 and Lemma 1 to the algorithm CGALG (introduced in [2]), reducing significantly the search space for finding an optimal solution. In particular, LGG-VSB avoids the construction of viewsets that do not rewrite the queries in the workload, by producing the candidate viewsets in such a way that the construction of the equivalent rewritings of the query is quickly achieved; i.e. instead of construction of every set of views whose body is a generalization of a subexpression of a query's body (CGALG), LGG-VSB constructs viewsets that form a partition of the body of each query in the workload.

Algorithm *LGG-VSB.*
   Input: A bag oriented view selection problem input[1] $\mathcal{P} = \{\mathcal{S}, \mathcal{Q}, \mathcal{D}, \mathcal{L}\}$.
   Output: Λ, the representative set of optimal solutions.
Begin
   1. Let $\mathcal{V}$ be a set of viewsets constructed as follows: Each $\mathcal{V}' \in \mathcal{V}$ is of
     the form $\mathcal{V}' = \mathcal{V}_1 \cup \ldots \cup \mathcal{V}_n$, where n is the number of queries in $\mathcal{Q}$
     and each viewset $\mathcal{V}_i$ is obtained from the query $Q_i \in \mathcal{Q}$ as follows:
       - Let P_i be a partition of the subgoals of Q_i.
       - For each block $B_j \in P_i$, add a view definition $V_{i,j}$ in $\mathcal{V}_i$ whose body
         consists of the atoms in B_j and whose head variables are the
         variables in $lvars(Q_i, B_j)$.
   2. Set $G_0 = \mathcal{V}$; set $i = 0$.
   3. **while** $G_i \neq \emptyset$ **do**
       - $G_{i+1} = \{\mathcal{V}_g | \mathcal{V}_g = (\mathcal{V}' - \mathcal{M}) \cup \{V_l\}$, where $\mathcal{V}' \in G_i$ and $\mathcal{M} \subseteq \mathcal{V}'$
         and $V_l = lgview(\mathcal{M})\}$.
       - i = i + 1.
   end while
   4. Let $\mathcal{V} = \bigcup_{j=0,\ldots,i} G_j$.
   5. Compute the cost $C(\mathcal{Q}, \mathcal{D})$ of $\mathcal{Q}$ on $\mathcal{D}$ and set it to C_{opt}.
   6. **For** every viewset $\mathcal{V}' \in \mathcal{V}$, such that $size(\mathcal{V}') \leq L$, **do**
       - Construct the set $\mathcal{R}_{\mathcal{V}'}$ of all equivalent rewritings of $\mathcal{Q}$ using $\mathcal{V}'$.
       - Set $\Lambda = \emptyset$.
       - **For** every distinct subset $\mathcal{R}$ of $\mathcal{R}_{\mathcal{V}'}$ such that $\mathcal{R}$ contains an
         equivalent rewriting of each query in $\mathcal{Q}$, **do**
           - Let $c = C(\mathcal{R}, \mathcal{V}'(\mathcal{D}))$.
           - **If** $c < C_{opt}$, **then** set $C_{opt} = c$ and set $\Lambda = \{(\mathcal{V}', \mathcal{R})\}$
           else if $c = C_{opt}$, **then** $\Lambda = \Lambda \cup \{(\mathcal{V}', \mathcal{R})\}$.
end.

5   Chain and Path Queries

In this section, we study the bag-oriented view selection problem when the query
workload is a set of either chain queries or path queries. The main results are as
follows: Subsection 5.1 demonstrates that for a problem input $\mathcal{P} = (\mathcal{S}, \mathcal{Q}, \mathcal{D}, \mathcal{L})$,
where $\mathcal{Q}$ is a workload of chain queries, we cannot restrict the space of optimal
solutions by searching admissible viewsets which contain only *chain-views*, i.e.
views defined by chain queries. Subsection 5.2 demonstrates that for a problem
input $\mathcal{P} = (\mathcal{S}, \mathcal{Q}, \mathcal{D}, \mathcal{L})$, where $\mathcal{Q}$ is a workload of path queries, if there exists
a solution for $\mathcal{P}$, then there is at least one optimal solution for $\mathcal{P}$ which is
constructed by an admissible viewset containing only path views (Theorem 2).

5.1   Chain-Query Workload

In this section we study the view selection problem for workloads containing
only chain-queries. In particular, we focus our attention on whether there is an

---
[1] Recall that $\mathcal{L} = \{L\}$, where L is a single storage limit constraint.

optimal solution constructed by a set of chain-views. Unfortunately, as the following proposition shows, there are cases in which none of the optimal solutions is constructed by a set of chain-views.

Proposition 5. *There exists at least one bag-oriented view selection problem input* $\mathcal{P} = (S, \mathcal{Q}, \mathcal{D}, \mathcal{L})$ *such that:*

- $\mathcal{Q}$ *is a set of chain queries, and*
- $\mathcal{P}$ *has optimal solutions but there is* **no** *optimal solution* $\Lambda = (\mathcal{V}, R)$ *such that* $\mathcal{V}$ *contains only chain queries.*

Proof. The following example proves this proposition.

Example 6. Consider a query workload $\mathcal{Q} = \{Q\}$ on a database schema $\mathcal{S}$ that contains the binary relations r_1, r_2 and r_3, where Q is the following chain query:

$$Q : q(X, Y) :\text{-} \ r_1(X, Z), r_2(Z, W), r_3(W, Y).$$

Consider also the following five viewsets $\mathcal{V}_i$, $i \in \{1, 2, 3, 4, 5\}$:

$\mathcal{V}_1 = \{V_{11}, V_{12}\}$, where:
$$V_{11} : v_{11}(X, Z, W, Y) :\text{-} \ r_1(X, Z), r_3(W, Y).$$
$$V_{12} : v_{12}(X, Y) :\text{-} \ r_2(X, Y).$$
$\mathcal{V}_2 = \{V_{21}, V_{22}\}$, where:
$$V_{21} : v_{21}(X, Y) :\text{-} \ r_1(X, Z), r_2(Z, Y).$$
$$V_{22} : v_{22}(X, Y) :\text{-} \ r_3(X, Y).$$
$\mathcal{V}_3 = \{V_{31}, V_{32}\}$, where:
$$V_{31} : v_{31}(X, Y) :\text{-} \ r_2(X, Z), r_3(Z, Y).$$
$$V_{32} : v_{32}(X, Y) :\text{-} \ r_1(X, Y).$$
$\mathcal{V}_4 = \{V_{41}\}$, where:
$$V_{41} : v_{41}(X, Y) :\text{-} \ r_1(X, Z), r_2(Z, W), r_3(W, Y).$$
$\mathcal{V}_5 = \{V_{51}, V_{52}, V_{53}\}$, where:
$$V_{51} : v_{51}(X, Y) :\text{-} \ r_1(X, Y).$$
$$V_{52} : v_{52}(X, Y) :\text{-} \ r_2(X, Y).$$
$$V_{53} : v_{53}(X, Y) :\text{-} \ r_3(X, Y).$$

Observe that the above viewsets are all possible viewsets constructed as described in Section 4.

Suppose that we are given database instance $\mathcal{D} = \{(r_1(\text{a,b}); 5), (r_2(\text{b,c}); 10), (r_3(\text{c,d}); 5)\}$. Considering a storage limit L=35 tuples, the following viewsets:

$$\mathcal{V}_1(\mathcal{D}) = \{(v_{11}(a, b, c, d); 25), (v_{12}(b, c); 10)\}$$
$$\mathcal{V}_5(\mathcal{D}) = \{(v_{51}(a, b); 5), (v_{52}(b, c); 10), (v_{53}(c, d); 5)\}$$

do not violate the storage limit constraint. In contrast, the viewsets:

$$\mathcal{V}_2(\mathcal{D}) = \{(v_{21}(a, c); 50), (v_{22}(c, d); 5)\}$$
$$\mathcal{V}_3(\mathcal{D}) = \{(v_{31}(a, c); 50), (v_{32}(c, d); 5)\}$$
$$\mathcal{V}_4(\mathcal{D}) = \{(v_{41}(a, c); 250)\}$$

do violate it. Thus, $\Lambda = (\mathcal{V}_1, R)$ and $\Lambda' = (\mathcal{V}_5, R')$ are solutions for input $\mathcal{P}$, where the rewritings R and R' are the following:

$$R : q(X, Y) :\text{-} v_{11}(X, Z, W, Y), v_{12}(Z, W).$$
$$R' : q(X, Y) :\text{-} v_{51}(X, Z), v_{52}(Z, W), v_{53}(W, Y).$$

Using the cost model presented in Section 3, the costs of Λ and Λ' are $C(R, \mathcal{V}_1(\mathcal{D})) = 55$ and $C(R', \mathcal{V}_4(\mathcal{D})) = 325$ respectively. As a consequence, Λ is an optimal solution for $\mathcal{P}$.

5.2   Path-Query Workload

In this section we study the view selection problem for *path-query workloads* (i.e. workloads of path queries). Unlike to the problem for chain query workloads in which we cannot reduce the search space to the class of chain views, for path-query workloads we can reduce the search space even more. The main result of this section, presented by the following theorem, is that whenever the workload is a set of path-queries, we can focus on *path-viewsets* whose views have at most as many subgoals as the length of the longest path-query in the workload.

Theorem 2. *Let $\mathcal{P} = (S, \mathcal{Q}, \mathcal{D}, \mathcal{L})$, be a conjunctive bag-oriented view selection input, and $\mathcal{Q}$ contains a set of path queries. If there exists a solution $\Lambda = (\mathcal{V}_o, \mathcal{R}_o)$ for $\mathcal{P}$, then there is an optimal solution $\Lambda' = (\mathcal{V}'_o, \mathcal{R}'_o)$ for $\mathcal{P}$ such that:*

- *each view in $\mathcal{V}'_o$ is defined as a path of the same relation as a query $Q \in \mathcal{Q}$,*
- *every view in $\mathcal{V}'_o$ has at most n subgoals, where n is the length of the longest query in $\mathcal{Q}$,*
- *every $R \in \mathcal{R}'_o$ is a chain query.*

Consequently, we may restrict our attention in searching optimal solutions constructed by path-viewsets. In this case, the number of admissible viewsets is exponential to the number of subgoals of the path-queries in the workload. This exponential bound is implied by the reduction of the problem of searching path-viewsets to the integer-partitioning problem [5].

Based on Theorem 2, we can improve the LGG-VSB for workloads containing only path-queries.In particular, when we know that the workload $\mathcal{Q}$ consists of n path-queries of the same relation, steps 1-4 of LGG-VSB can be replaced by:

- Each $\mathcal{V}_{\mathcal{I}} \in \mathcal{V}$ contains a path-view V_k of length k, for every distinct integer $k \in \mathcal{I}$, where the set of integers $\mathcal{I}$ is of the form $\mathcal{I} = \mathcal{I}_{k_1} \cup \ldots \cup \mathcal{I}_{k_n}$, and $\mathcal{I}_{k_i}$ is a partition of the length of path-query $P_{k_i} \in \mathcal{Q}$, $i \in \{1, \ldots n\}$; the partitions of an integer can be computed using an algorithm from [26].

6   Conclusion

In this paper we studied the problem of view selection under bag semantics. In particular, we investigated ways to limit the search space of candidate views, given a workload of CQs. We improved previous results by exploiting very refined characterizations of views that participate in equivalent rewritings. Based on these characterizations we proposed sound and complete algorithms to select views for a query workload. Besides, we studied the problem in two special cases, that is, when the workload contains only (a) chain queries, or (b) path queries, and present interesting results which further improve the proposed algorithm. Concerning the experimental evaluation of our approach, we have contacted preliminary experiments that gave promising results.

There is a lot to be done for future work including the following: (a) studying further the potential features of lgviews, (b) studying more special cases of the view selection problem, (c) studying the view selection problem for parameterized queries, and (d) studying the exact complexity of the problem.

Acknowledgements. We would like to thank Timos Sellis and the anonymous reviewers for their valuable comments.

References

1. Abiteboul, S., Hull, R., Vianu, V.: Foundations of Databases. Addison-Wesley, Reading (1995)
2. Afrati, F., Chirkova, R., Gergatsoulis, M., Pavlaki, V.: View selection for real conjunctive queries. Acta Inf. 44(5), 289–321 (2007)
3. Afrati, F.N., Chirkova, R.: Selecting and using views to compute aggregate queries (extended abstract). In: Eiter, T., Libkin, L. (eds.) ICDT 2005. LNCS, vol. 3363, pp. 383–397. Springer, Heidelberg (2004)
4. Afrati, F.N., Li, C., Ullman, J.D.: Generating efficient plans for queries using views. In: SIGMOD Conference 2001, pp. 319–330 (2001)
5. Andrews, G.E., Eriksson, K.: Integer Partitions. Cambridge University Press, Cambridge (2004)
6. Baralis, E., Paraboschi, S., Teniente, E.: Materialized views selection in a multidimensional database. In: VLDB 1997, pp. 156–165 (1997)
7. Chirkova, R., Genesereth, M.R.: Linearly bounded reformulations of conjunctive databases. In: Palamidessi, C., Moniz Pereira, L., Lloyd, J.W., Dahl, V., Furbach, U., Kerber, M., Lau, K.-K., Sagiv, Y., Stuckey, P.J. (eds.) CL 2000. LNCS, vol. 1861, pp. 987–1001. Springer, Heidelberg (2000)
8. Chirkova, R., Halevy, A.Y., Suciu, D.: A formal perspective on the view selection problem. The VLDB Journal 11(3), 216–237 (2002)
9. Chirkova, R., Li, C.: Materializing views with minimal size to answer queries. In: PODS, pp. 38–48 (2003)
10. Florescu, D., Levy, A.Y., Suciu, D., Yagoub, K.: Optimization of run-time management of data intensive web-sites. In: VLDB 1999, pp. 627–638 (1999)
11. Gupta, H., Harinarayan, V., Rajaraman, A., Ullman, J.D.: Index selection for OLAP. In: ICDE 1997, pp. 208–219 (1997)
12. Gupta, H., Mumick, I.S.: Selection of views to materialize in a data warehouse. IEEE Trans. Knowl. Data Eng. 17(1), 24–43 (2005)
13. Harinarayan, V., Rajaraman, A., Ullman, J.D.: Implementing data cubes efficiently. SIGMOD Rec. 25(2), 205–216 (1996)
14. Karloff, H., Mihail, M.: On the complexity of the view-selection problem. In: PODS 1999, pp. 167–173 (1999)

15. Lloyd, J.W.: Foundations of logic programming. Springer, Heidelberg (1984)
16. Plotkin, G.: A note on inductive generalization. Machine Intelligence 5, 153–163 (1970)
17. Pottinger, R., Halevy, A.: Minicon: A scalable algorithm for answering queries using views. The VLDB Journal 10(2-3), 182–198 (2001)
18. Rizzi, S., Saltarelli, E.: View materialization vs. indexing: Balancing space constraints in data warehouse design. In: Eder, J., Missikoff, M. (eds.) CAiSE 2003. LNCS, vol. 2681, pp. 502–519. Springer, Heidelberg (2003)
19. Surajit Chaudhuri, M., Vardi, M.Y.: Optimization of real conjunctive queries. In: PODS 1993, pp. 59–70 (1993)
20. Theodoratos, D., Sellis, T.K.: Data warehouse configuration. In: VLDB 1997, pp. 126–135 (1997)
21. Theodoratos, D., Xu, W.: Constructing search spaces for materialized view selection. In: DOLAP, pp. 112–121 (2004)
22. Ullman, J.D., Garcia-Molina, H., Widom, J.: Database Systems: The Complete Book. Prentice Hall PTR, Upper Saddle River (2001)
23. Xu, W., Theodoratos, D., Zuzarte, C.: Computing closest common subexpressions for view selection problems. In: DOLAP, pp. 75–82 (2006)
24. Yu, J.X., Choi, C.-H., Gou, G., Lu, H.: Selecting views with maintenance cost constraints: Issues, heuristics and performance. Journal of Research and Practice in Information Technology 36(2), 89–110 (2004)
25. Zhou, J., Larson, P.-A., Freytag, J.C., Lehner, W.: Efficient exploitation of similar subexpressions for query processing. In: SIGMOD Conference, pp. 533–544 (2007)
26. Zoghbi, A., Stojmenović, I.: Fast algorithms for generating integer partitions. Int. J. Comput. Math. 70(2), 319–332 (1998)

QoS-Aware Publish-Subscribe Service for Real-Time Data Acquisition*

Xinjie Lu[1,4], Xin Li[3], Tian Yang[1,4], Zaifei Liao[1,4],
Wei Liu[1], and Hongan Wang[1,2]

[1] Institute of Software, Chinese Academy of Sciences, Beijing 100190, China
xinjie05@ios.cn
[2] State Key Lab. of Computer Science, Institute of Software,
Chinese Academy of Sciences, Beijing 100190, China
[3] Department of Computer Science and Technology, Shandong University,
Jinan Shandong 250101, China
[4] Graduate University of the Chinese Academy of Sciences, Beijing 100049, China

Abstract. Many complex distributed real-time applications need complicated processing and sharing of an extensive amount of data under critical timing constraints. In this paper, we present a comprehensive overview of the Data Distribution Service standard (DDS) and describe its QoS features for developing real-time applications. An overview of an active real-time database (ARTDB) named Agilor is also provided. For efficient expressing QoS policy in Agilor, a Real-time ECA (RECA) rule model is presented based on common ECA rule. And then we propose a novel QoS-aware Real-Time Publish-Subscribe (QRTPS) service compatible to DDS for distributed real-time data acquisition. Furthermore, QRTPS is implemented on Agilor by using objects and RECA rules in Agilor. To illustrate the benefits of QRTPS for real-time data acquisition, an example application is presented.

Keywords: QoS, Real-Time Publish-Subscribe, ECA Rule, Active Real-Time Database.

1 Introduction

Many complex distributed real-time applications require complicated processing and sharing of an extensive amount of data under critical timing constraints. These applications include collecting data from the environment, processing acquired data in the context of historical data and providing timely response. How to transmit or disseminate these data timely and exactly by simple configuration is a noticeable problem as yet.

Image that we periodically receive sensor data from a mine in a colliery, these data might contain information about gas, temperature, smog, etc. of every *Observation Point*. The system is required to disseminate these data under specific

---

* This work was supported in part by the National High Technology Research and Development Program ("863"Program) of China under Grant No. 2006AA04Z182.

M. Castellanos, U. Dayal, and T. Sellis (Eds.): BIRTE 2008, LNBIP 27, pp. 29–44, 2009.

timing constraints for the following abnormal scene detection. Hereinto, publish-subscribe model is very suitable for such repetitive, time-critical data distribution. A limitation of most existing architectures that support publish-subscribe is their limited support for the expression and enforcement of Quality of Service (QoS) parameters (such as required bandwidth or latency, for instance). This observation ranges from models, such as the CORBA Event Service [2], CORBA Notification Service [1], Java Message Service [3], to systems, such as CEA (Cambridge Event Architecture) [4], Distributed Asynchronous Collections [9], SIENA (Scalable Internet Event Notification Architectures) [8] or Cayuga [27]. This is a significant shortcoming, since QoS features are an important component of applications, and they have been widely studied in the context of direct communication [5,6,7,10,22,24,25].

Data Distribution Service (*DDS*) is a newly adopted specification from the Object Management Group (OMG). *DDS* is aimed at a diverse community of users requiring data-centric publish-subscribe communications. *DDS* departs from previous approaches in two primary aspects: (1) enumerating and providing formal definitions for the QoS settings that can be used to configure the service, and (2) the tight binding of a "topic" to a data-type, along-with the additional QoS settings, implementing optimizations such as pre-allocating the resources needed to send or receive a "topic" [11].

To meet the requirements of real-time and active capabilities described in QoS policies of *DDS* [15], we introduce active real-time databases (ARTDB) [16] to implement these QoS policies. The ARTDB [17] is proposed to provide both active and real-time capabilities. In the context of an ARTDB, data distribution can be implemented via ECA rules, and applications can consume data at specific rate on specific condition. So it would be desirable to develop a QoS-aware Real-Time Publish-Subscribe (QRTPS) system on ARTDB. This system infrastructure should be efficient, scalable, flexible, and cater for the architecture of active real-time database system, except for providing real-time predictability. The contributions of this paper are:

1. An original and highly flexible real-time publish/subscribe system, QRTPS, that supports QoS in the subscription and the publication. It supports QoS policies settings and is fully implemented with *Agilor*. The QRTPS allows for various configurations to express different users' QoS requests only by minor programming effort, whereas in traditional distributed systems providing so many QoS features is an error prone and complex task.
2. For expressing QoS policies effectively and setting conveniently, we propose a Real-time ECA rule model (*RECA*) that extends common ECA in complicated temporal events, composite conditions and several coupling models. By means of the primitives defined in *Agilor*, all of QoS policies can be conveniently configured for both subscription and publication through *RECA*.
3. In order to illustrate some features of QRTPS, a simplified example application is presented, which is a sensor-based active monitoring system. The related data structures, parameters settings of QoS and translated *RECA* rules are described in detail.

The remainder of this paper is organized as follows. The next section introduces high-level design and active object model of *Agilor* as background, for further incorporating QRTPS into *Agilor*. In Section 3, we propose Real-time ECA (*RECA*) rule model to express QRTPS. Section 4 discusses a simplified example application to illustrate some of the previously mentioned features of QRTPS. We conclude this paper and present future work in Section 5.

2   Background

2.1   Overview of *Agilor*

Agilor is a typical active real-time database and its architecture as Figure 1 consists of some kernel modules and critical services. We present the main function of each component in sequence and introduce an example to illustrate the operation mechanism of *Agilor*.

Key Components and Their Duties: The *Storage Manager* takes charge of persistent objects and rules storage on disks and support read/write interfaces. The *Object Manager* and *Rule Manager*, resident in main-memory, are responsible to add/delete/update objects and ECA rules, respectively. The *Transaction Scheduler* deals with all transactions and access objects through interface in the *Object Manager*.

The *Rule Manager* not only stores rules into the rule-base, but also performs rule processing by the *Event Detector* and *Condition Evaluator*. The *Rule Manager* also submits actions and necessary parameters (e.g. deadline, worst-case execution time) to the *Transaction Scheduler*. The *Event Detector* monitors events occurred in database and system. The *Condition Evaluator* checks whether specific conditions are satisfied on receiving events from *Event*

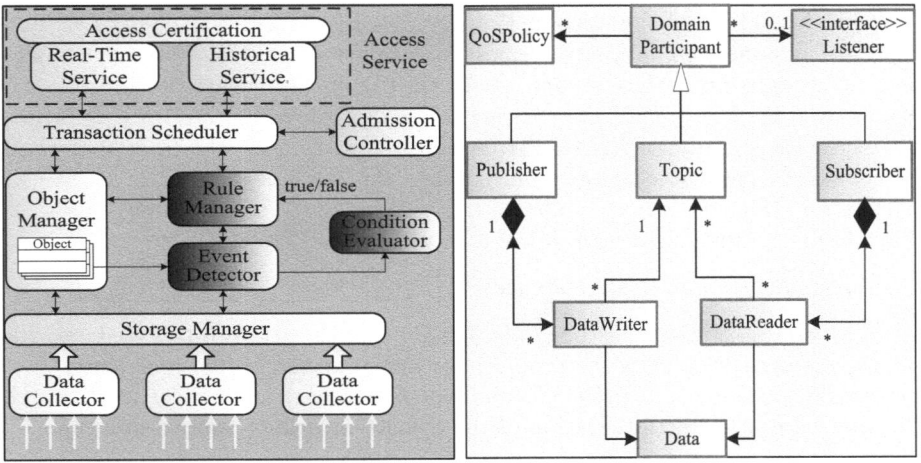

Fig. 1. *Agilor* Architecture   **Fig. 2.** *DCPS* conceptual model [12,23]

Detector. When conditions are satisfied, relevant actions will be submitted by *Rule Manager* to *Transaction Scheduler* to execute.

The *Admission Controller* ensures that admitted transactions do not over-burden the system. This module inspects whether to accept or reject a new transaction based on a feedback mechanism considering resources and workload in system and the importance of the transaction.

The data access services consist of *Real-time Service, Historical Service* and *Access Certification.* They support retrieval of historical data as well as real-time data (synchronous mode or subscription mode) under time constraints. In these services, real-time publish-subscription service provides push mechanism based on *RECA* rules. The *access certification service* ensures that access is provided only to entitled applications. An important building block of the *Agilor* is ECA rule and we will discuss its extension edition Real-time ECA in Section 3.

Real-time Data Acquisition Example: We assume that a real-time production monitoring system needs instantaneous flow readings from a production device *PD* timely. And it also need to accumulate the instantaneous readings to form another data, named accumulated readings.

This requirement can be divided into two asynchronous flow: *Collecting* and *Querying. Collecting* finishes getting instantaneous readings from *PD*, saving them in a certain topic(refer to *INS*) and accumulating them to another topic(refer to *ACC*) in *Agilor. Querying* achieves continuous querying the latest value in these two topics.

Collecting: Real-time instantaneous readings are collected by *Data Collector* and then stored in *INS* by *Storage Manager*. Once *INS* are updated, *Event Detector* will find this change and trigger *Condition Evaluator* to evaluate whether this new instantaneous reading is legal. If the result is true, *Rule Manager* will submit the accumulation action to *Transaction Scheduler*. Then, *Transaction Scheduler* will execute this action in appropriate occasion. As a consequence, *ACC* are also updated timely.

Querying: The real-time production monitoring system will query *INS* and *ACC* every second. The query request is submitted to *Real-time Service* and passes through the authentication of *Access Certification* with IP address and Application ID. Then, *Real-time Service* sends this request to *Transaction Scheduler* with the permission of *Admission Controller*. Finally, the latest values in *INS* and *ACC* return as the response of that query request.

2.2   Conceptual Model of *DDS*

DDS describes two levels of interfaces [12]:

- A lower *DCPS (Data-Centric Publish-Subscribe)* level that is targeted towards the efficient delivery of the proper information to the proper recipients.
- An optional higher *DLRL (Data Local Reconstruction Layer)* level, which allows for a simple integration of the Service into the application layer.

We restrict our discussion of *DDS* to the *DCPS* layer. The overall *DCPS* model is illustrated in Figure 2, which consists of the following entities: *DataWriter,*

Table 1. *DDS* QoS Policies [19]

QoS Policy	Meanings
Durability	Determines if data outlives the time when written or read.
Deadline	Determines rate at which periodic data is refreshed.
Latency_Budget	Sets guidelines for acceptable end-to-end delays.
Ownership	Controls *writer(s)* of data.
Ownership_Strength	Sets ownership of data.
Transport_Priority	Allows the application to take advantage of transports capable of sending messages with different priorities
Liveliness	Sets liveness properties of *topics, data readers, data writers*.
Time_Based_Filter	Mediates exchanges between consumers and producers.
Reliability	Controls reliability of data transmission.
History	Sets how much data is kept to be read.
Resource_Limits	Controls resources used to meet requirements.

DataReader, Publisher, Subscriber, and *Topic.* All these classes extend *Domain-Participant*, representing their ability to be configured through QoS policies and each of them has a set of QoS Policies that are suitable to it.

2.3   Supported QoS of *DDS*

The *DCPS* entities in *DDS* include *Topics*, which describe the type of data to be writ-ten or read; *Data Readers*, which subscribe to the values or instances of particular topics; and *Data Writers*, which publish values or instances for particular topics. Various properties of these entities can be configured using combinations of the 22 QoS policies. Moreover, *Publishers* manage groups of *data writers* and *Subscribers* manage groups of *data readers*. We summarize all the *DDS* QoS policies related to our work in Table 1 and a more detailed discussion can be found in [12]. Each QoS policy has several attributes with the majority of the attributes having a large number of possible values, e.g. an attribute of type long or character string. Moreover, not all QoS policies are applicable to all *DCPS* entities, nor are all combinations of policy values semantically compatible [19,23].

3   Real-Time ECA

ECA Rules are used to specify constraints that define correct states of objects as well as actions to be taken on certain events. To efficiently describe QoS policy in *Agilor*, a Real-time ECA (*RECA*) rule model is proposed, which extends common ECA [21] in complicated temporal events, composite conditions and coupling models. The *RECA* rule model is divided into three parts: **Event**, **Condition** and **Action**.

Formally, a rule is modeled by $<RN, RD, RV, E, C, A, CMEC, CMCA, CMEA, CMRR>$, in which *RN, RD, RV* is the name, deadline and value of the rule, respectively. The deadline reflects urgency of the rule (including the action) while the value reflects importance of the rule. The value decides the order of condition evaluations when multiple rules are triggered at the same time. E is a set of events that can invoke rules, C is a set of conditions and A is a set of ordered actions. Actions should be taken when specific conditions are satisfied. *CMEC* and *CMCA* are the coupling modes between event and condition evaluation and between condition and action execution separately, i.e. when condition evaluation and action execution can take place relative to the time of triggering events. *CMEA* and *CMRR* are the coupling modes between event and action execution and between two rules [20,21].

3.1   Event

Events are occurrences of interests which are predefined in the system such as data update events and clock events. Events can be divided into primitive events, which refer to simple and atomic events, and composite events, which consist of primitive events combined with event operators.

Events can be described formally as follow:

$$
\begin{aligned}
E ::= {}& p|(\neg E)|(O_{[t1,t2]}E)|(E_1{\wedge}E_2)|(E_1{\vee}E_2)|(E_1 O_{[t1,t2]}E_2)|(E_1 O_{\leq t2}E_2) \\
& |(O_{[t1,t2]}(E_1{\rightarrow}E_2))|(O_{[t1,t2]}(E_1{\wedge}E_2))|(O_{[t1,t2]}(E_1{\vee}E_2)) \\
& |(O_{[t1,t2]}((E_1{\rightarrow}E){\wedge}(E{\rightarrow}E_2))).
\end{aligned} \tag{1}
$$

The predication p is a primitive event. The symbol '$\neg$','$\wedge$'and'$\vee$' stand for negation(not), conjunction(and), disjunction(or) operator, respectively. $(E_1 O_{[t1,t2]}E_2)$ is a sequence of events to occur over a time interval [t1,t2], in which the latter event E2 must occur after the former event E1 between [t1,t2]. The composite event arises when the last event in the sequence has occurred.

Three kinds of primitive events are realized in *Agilor* and they are

1. *system events*: some particular events of operating system or database system, e.g. OnTimer and OnIOError;
2. *method events*: data manipulation events of an object/class, e.g. OnUpdate and OnDelete;
3. *custom events*: events predefined by user's application for specified purpose, such as sensor failure event. Custom events are always triggered explicitly by application calling RaiseCustomEvent() function.

The *method events* are primary events in ARTDB and any data manipulation event of an object/class can be a potential *method event*. *Method events* will be triggered automatically when the corresponding method is invoked.

A method event can be defined by 6-tuple <*EN, T, OM, CMME, EP, SL*>, where *EN* is the name of event, *T* is the time of occurrence, *OM* is the name of object method which should be one of the existing methods in object base, and the coupling mode *CMME* is an indication of whether the event should be generated before or after the execution of the method. *EP* is the parameter set

Table 2. Extension of Complicated Temporal Events

Event type	Definition	Semantics
Durative event	event E occurs at regular intervals between two time instants, X and Y.	$O_{[X,Y]}E$
Time constrained sequence	E1 Seq-Within[X seconds] E2, occurs when both E1 and E2 have occurred in that order within X seconds.	$E_1 \rightarrow O_{\leq X} E_2$
Durative sequence	E1 Seq-During[X,Y] E2, occurs when both E1 and E2 have occurred in that order at regular intervals from time instant X to Y.	$O_{[X,Y]}(E_1 \rightarrow \Diamond E_2)$
Durative conjunction	E1 AND-During[X,Y] E2, occurs when both E1 and E2 have occurred in any order at regular intervals from time instant X to Y.	$O_{[X,Y]}(E_1 \wedge E_2)$
Durative disconjunction	E1 OR-During[X,Y] E2, occurs when either E1 or E2 occurs or when both E1 and E2 occur at regular intervals from time instant X to Y.	$O_{[X,Y]}(E_1 \vee E_2)$
Durative between	Between-During (E1, E2)[X,Y], occurs when there are events occur between event E1 and E2 between starting time X and ending time Y, ignoring the relative order of their occurrences.	$O_{[X,Y]}((E_1 \rightarrow \Diamond E)$ $\wedge (E \rightarrow \Diamond E_2))$

of the event corresponding to the parameters of the method which will be passed to condition evaluator for check. *SL* is a subscribers list made up of the rules and composite events which subscribe this event. The subscribers of this event will be notified when the event occurs.

In order to express more complicated temporal events [21], we extend common ECA rules as shown in Table 2. The six kinds of patterns focus on duration-related aspect of complicated event and all have typical application scenarios in real-world. For the sake of limited space, we give two examples. We still use the data acquisition scene introduced in Section 2.1. Let E1 denote the event of update on topic *INS* and E2 denote the event of update on topic *ACC*. We define t as an arbitrary time instant and n as an integer. *Durative Event* event, $O_{[t,t+n]}E1$, can occur every n second because E1 occurs per second between t and t+n. *Durative Sequence* event, E1 Seq-During[t,t+n] E2, can occur every n seconds because both E1 and E2 have occurred in that order every second.

For complicated temporal events scan, a useful approach has been to adopt Nondeterministic Finite Automata(NFA) to represent the structure of an event sequence [26]. Furthermore, the NFA-based approach can be extended to handle sequence construction, as proposed in YFilter [28] in the context of XML message filtering.

3.2   Condition

The event indicates the need to check; whereas the condition determines what to check. The condition set C describes the situations that are used to check whether all prerequisites are satisfied for actions.

Conditions can be described formally as follow:

$$C ::= p|(\neg C)|(C_1 \wedge C_2)|(C_1 \vee C_2). \tag{2}$$

The predication p is a primitive condition and in *Agilor* it can be

(1) *Selection condition*: evaluation of a single attribute value of one object (e.g. OP.Gas>20),
(2) *Aggregation condition*: comparison of a single attribute aggregated over multiple instances (e.g. Max(OP.Smog)>100),
(3) *Join condition*: comparison of a single common attribute of multiple homogeneous objects (e.g. OP1.Pressure=OP2.Pressure),
(4) *Transition condition*: comparison of a single attribute over multiple instances (e.g. OP1.Gas>OP1.GetLast(Gas)) and
(5) *Application-specific condition*: evaluation of functions predefined by applications.

Applications also can define composite conditions by combining a set of primitive conditions with logical operators such as disjunction and conjunction.

3.3   Action

The action set A defines a set of ordered actions, which are similar to the definition of methods in object model. Actions could be database operations including deletion and update, as well as external actions such as procedure call (e.g. publishing data or signaling an alarm to applications). Deadline as an additional parameter should be assigned to the action. It is a relative delay to the occurrence time of the triggering event. For example, a triggered action must be finished in 10 milliseconds after a temporal attribute X is updated.

3.4   Coupling Modes

The *CMEC, CMCA, CMEA* and *CMRR* identify the time semantics when condition evaluation and action execution can take place relative to the triggering event, with the constraint that condition evaluation must be performed before action execution. The optional values and meanings of each coupling mode are summarized in Table 3. If *CMEC* is configured to *immediate*, when event occurs, the current running transaction will be suspended, and condition evaluation is performed immediately. While *detached*, the evaluation of condition will be finished in a different transaction. Similarly, *CMCA* defines such a relationship between condition and action. *CMEA* describes this relationship between event and action. *CMEC, CMCA, CMEA* are aimed at dealing with each part in one ECA rule, while *CMRR* focuses on the relationship between ECA rules. Using

Table 3. Coupling modes definition

Name	Optional Value	Meanings
CMEC	immediate	When events occur, the transaction is suspended,and condition evaluation is performed immediately.
	detached	Condition evaluation is done in another transaction.
CMCA	immediate	The triggered action is executed immediately after condition evaluation.
	detached	The triggered action is treated as a new separate transaction.
CMEA	immediate	The triggered action is executed immediately after event occurs.
	detached	The triggered action is treated as a new separate transaction.
CMRR	immediate	Execution of one rule immediately triggers another rule.
	concurrent	Many rules may be triggered at the same time.

these two optional values, *immediate* and *concurrent*, we can build a ECA rules chain or a ECA rules net to achieve more sophisticated business flow.

To avoid unpredictable increase of the execution time of the triggering transaction in a real-time environment, the combinations of *CMEC* and *CMCA* had better be *immediate-detached* or *detached-immediate*. Similarly, the limit is put on the depth of triggered rules to avoid uncontrolled cascade triggering.

3.5 Semantic for *RECA* Rules

The basic structure of the rules in *Agilor* is expressed as triggering events, conditions and actions, as well as the timing constraints and coupling modes. Deadline and value are considered in each *RECA* rule and the semantic for rules is defined as follow:

Rule::=BEGIN RULE *<RuleName>*
  VALUE *<Value>*
  WHEN *<Event>*
  IF *<Condition>* **CMEC** *[immediate|detached]*
        CMEA *[immediate|detached]*
  THEN   *<Action>* **WITHIN** *<Deadline>*
  CMCA *[immediate|detached]*
  CMRR *[immediate|concurrent]*
END RULE

4   Case Study

This section aims to illustrate some of features of QRTPS with a simplified example application. It is developed by the commercially available QRTPS on

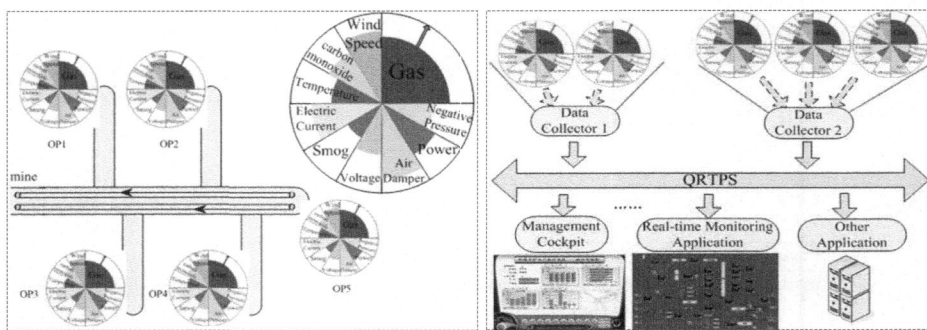

Fig. 3. Overview of the example application

Agilor. The example uses QRTPS for the implementation of a sensor-based active monitoring system. An overview of the proposed system is shown in Figure 3. The system consists of five *Observation Points(OP)* in a coal mine and there are many sensors on each *OP* to measure different indicators. For the sake of simplicity, we consider three sensors (e.g. gas, temperature, smog) on each *OP*.

We use the following requirements for the overall system to demonstrate the most important features of QRTPS:

1. **Fault Tolerant:** The data collector and connected applications only consider the temperature readings of *OP1* at a time. If no data from this most-trusted sensor is received within 5 second, temperature data shall automatically be received from another *OP*, thus allowing a seamless failover [29].
2. **Composite Event:** When the gas readings keep increasing rapidly and temperature readings are greater than a threshold value, the transport priorities and frequencies of them should be enhanced. Transport priorities of gas and temperature readings shall be set to 100 and monitoring application shall receive gas and temperature readings every second.
3. **Dynamic Resource Allocation:** At the start-up of a production device, Real-time Monitoring Application shall dynamically create a subscription about temperature of the equipment and receive data every 5 seconds.

4.1   Related Data Structures

In order to decouple sensor readings and their meta information, we define two data structures [29], which shall be exchanged between the components, for each sensor type. The first data structure contains the actual sensor readings to be transferred. Additional meta information are modeled in an additional structure and only need to be published when the application starts or if sensors are exchanged. Thus, subscription can be dynamically reconfigured to accurately interpret incoming data from different sensors. As an example, the class definitions for gas sensors are shown below. The class *Gas* is used to transmit the sensor readings, whereas *GasSensorInfo* contains meta-information to interpret the sensor data correctly.

Class Gas{	Class GasSensorInfo{

private:
    long   datacollector_id;
    long   OP_id;
    double   value;
};
	private:
	    MeasuringUnit unit;
	    double maxGas;
	    double minGas;
	    double sampleRate;
	    double thresholdValue;
	    double abnormalRate;
	};

4.2 QoS Policies Settings of Each Entity

In Figure 4, p1 to p5 describe the steps in a publication and s1 to s5 express the processes in a subscription. QoS policies settings for each requirement above mentioned are shown in Table 4.

The p1 and p2 of Figure 4 show the creation of the *Publisher* and *DataWriter* respectively. The p3 and p4 show the process of user application writing data to *DataWriter* and the process of *DataWriter* writing data to *Topic Queue*. The p5 step shows that the corresponding notifications are propagated according to the current *Publisher*'s policy on sending.

The s1 of Figure 4 shows the *Subscriber*'s creation. The s2 shows the life cycle of a *SubscriberListener*. Firstly, it must be created and attached to the *Subscriber*. Then when notification arrives, it is made available to each related *DataReader*. After that, the *SubscriberListener* is triggered (s3). The application must get the list of affected *DataReader* objects; then it can read (s4) the data directly from *Topic Queue*. The s5 shows the process of user application reading data from *DataReader*.

The first requirement describes that subscribing applications shall get temperature readings from the second-most-trusted sensor, when the most-trusted sensor stops working because of damage or other reasons. We set the OWN-ERSHIP.kind parameter to "exclusive" to ensure that readers will only receive data from a single sensor. The DEADLINE policy defines the timeout that the subscribers will automatically failover to the second-most-trusted sensor. Thus, fault-tolerant distributed applications can easily be developed with the ability to dynamically react to failures in the system [29].

The second requirement allows enhancing the transport priorities and frequencies during run-time. When the gas readings grow too rapidly and temperature readings are greater than a threshold value, some anomalies are likely to happen. To ensure that monitoring application receives these real-time data with higher frequency and reliability, we can set the TIME_BASED_FILTER policy to 1 to guarantee *DataReader* reading every second. Since the TRANSPORT_PRIORITY of *DataWriter* is set to 100, it is assured that *DataWriter* will give first priority to

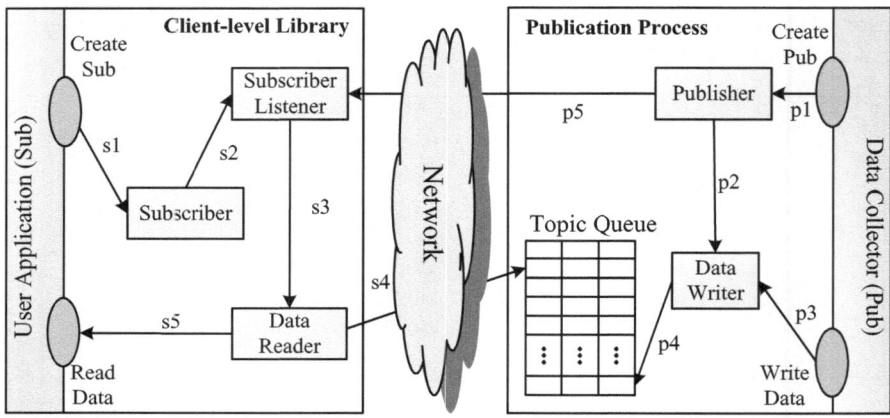

Fig. 4. Framework and Sequence of QRTPS Service

Table 4. QoS Settings for this Case

	Data Entity	Policy of IssueConsumer	Policy of IssueProducer
1	Temperature	OWNERSHIP.kind=EXCLUSIVE; OWNERSHIP_STRENGTH.value= datacollectorid of the data collector; DEADLINE.period.sec = 5;	OWNERSHIP.kind =EXCLUSIVE; DEAD- LINE.period.sec = 5;
2	Gas, Temperature	TIME_BASED_FILTER =1s; TRANSPORT_PRIORITY =100;	
3	Temperature	LIVELINESS.kind =MANUAL_BY_TOPIC; LIVELINESS.lease_duration =infinite; TIME_BASED_FILTER =5s;	LIVELINESS.kind =MANUAL_BY_TOPIC; LIVELINESS.lease_duration =infinite;

these emergent data. In this way, composite event based dynamic priority mechanism can be implemented with the configurations of QoS policies.

The third requirement needs to dynamically create a subscription and receive data every 5 seconds, when a production device starts up. To create a subscription at random time, we can set the LIVELINESS.kind=MANUAL_BY_TOPIC and LIVELINESS.lease_duration=infinite for both *DataReader* and *DataWriter*. Then, TIME_BASED_FILTER of temperature is set to 5 to make sure that *DataReader* can receive temperature readings every 5 seconds, regardless how fast the *DataWriter* publish this information. Similarly, we can dispose these resources used by this subscription when the equipment stops. By this means, dynamic resource allocation is achieved with emphasis on resource management on the grounds of actual production condition.

```
BEGIN RULE Rule_1    //Req. 1
  VALUE  1
  WHEN OnTimer("Timer1",10)
  IF True  THEN
      variant currentvalue;
      long currenttime;
      long begintime;
      begintime = CurrentTime();
      while(TopicState("TempSensor2")) {
          currentvalue =TopicValue("OP1.TempSensor1",1);
          currenttime = CurrentTime();
          if((currenttime- begintime)>5) {
              currentvalue =TopicValue("OP2.TempSensor2",1);

              ......
              break;
          }
      }
  WITHIN 3
END RULE
```

```
BEGIN RULE Rule_2 // Req. 2
  VALUE  1
  WHEN OnTimer("Timer2 ",5)
  GasMetaInfo ggasmetainfo;
  IF (TopicValue("OP1.GasSensor1",1)-TopicValue("OP1.GasSensor1",2))>
        ggasmetainfo.abnormalRate
    and  TopicValue("OP1.TempSensor3",1)> ggasmetainfo.thresholdValue
  THEN
    SetTimer("Timer3",1);

  ......
  WITHIN 3
END RULE
```

```
BEGIN RULE Rule_3.1 // Req.3          BEGIN RULE Rule_3.2 // Req.3
  VALUE  1                              VALUE  1
  WHEN OnStartUp("DeviceID")            WHEN OnTimer("Timer4")
  IF True  THEN                         IF True  THEN
    SetTimer("Timer4",5);                 variant currentvalue;
    EnableRule("Rule_3.2");               currentvalue=
  WITHIN 3                                    TopicValue("OP1.TempSensor1",1);
  CMRR [immediate] Rule_3.2             ......
END RULE                                WITHIN 3
                                      END RULE
```

Fig. 5. *RECA* Rules for Requirements

4.3   Translating QoS Settings to *RECA*

Each requirement's QoS parameters can be translated into a *RECA* rule or a set of *RECA* rules coupled by *RECA*'s coupling modes (Section3.4). Due to space limit, we translate QoS parameters only mentioned on *DataReader* in Table 4 into *RECA* rules as shown in Figure 5.

Primitives, defined in *Agilor*, provide many primary functions for *RECA* rules. For example, TopicState("TempSensor2") in Rule_1 is to check the running state of a Topic on TempSensor2 and return true only if it runs normally. TopicValue (*TopicName,index*) gets the value identified by *index* from a Topic namely *TopicName*. CurrentTime() will get the execution time of this rule. To achieve triggering some actions at regular intervals, OnTimer*(TimerName,interval)* is defined to be triggered every *interval* seconds. While SetTimer(*TimerName, interval*) is used to set the interval of a timer namely *TimerName*. To implement the coupling modes between rules, EnableRule(*RuleName*) and DisableRule(*RuleName*) are defined to make a rule activated or deactivated, respectively.

4.4   Limitations of Triggered Rules

Although using event-driven rules(especially ECA rules) has several benefits(e.g. reactivity, flexibility and manageability), the method also has some limitations. We present the limitations we have obtained during the study of using ECA for QoS management [30].

- ECA rules are more suitable for procedural programming, not object-oriented programming. As we mentioned in Section 4.1, some data structures defined by object-oriented approach, but we have to map these objects to property-value pairs during implementation.
- ECA rules do not have states, which means they can not record the progress of current instance. The common way of compensation is saving some intermediate results in a relational or real-time database for potential use in the following rules.
- Visualization of ECA rules is not easy to be implemented because only visualizing single ECA rule is not enough. It is more important to visualize the entire ECA rules chain(or net) which can achieve a whole business flow. In addition, the validation of ECA rules chain(or net) also n eeds visualization of these rules.

5   Conclusions and Future Work

In this paper, a new *DDS* compatible real-time service for data-centric publish-subscribe communication has been presented. The service is particularly targeting real-time applications which needs to manage resource consumption and timeliness of the data transfer. QRTPS allows many-to-many communication and alleviates a number of common problems which are of particular interest for the development of distributed assembly systems. For example, with its complex QoS support, fault-tolerant service, composite event based dynamic priority mechanism and dynamic resource allocation are achieved automatically and in an efficient manner. Future work can focus on adaptive adjusting QoS parameters during run-time in order to provide better performance with limited resources.

Acknowledgments. We would like to thank the anonymous referees for many valuable comments and suggestions, which helped improve the quality of this paper.

References

1. Object Management Group, OMG Headquarters, 250 First Avenue, Suite 201, Needham, MA 02494, USA. Notification Service Specification (2000)
2. Object Management Group, OMG Headquarters, 250 First Avenue, Suite 201, Needham, MA 02494, USA. Event Service Specification (2001)
3. Sun Microsystems, 901 San Antonio Road, Palo Alto, CA 94303, USA. Java Message Service (1999)
4. Bacon, J., Moody, K., Bates, J., Hayton, R., Ma, C., McNeil, A., Seidel, O., Spiteri, M.: Generic support for distributed applications. IEEE Computer, 68–76 (2000)
5. Blake, S., Black, D., Carlson, M., Davies, E., Wang, Z., Weiss, W.: An architecture for differenciated services, RFC 2475 (1998)
6. Braden, E.R., Zhang, L., Berson, S., Herzog, S., Jamin, S.: Resource reservation protocol (RSVP)-version 1 functional specification, RFC 2205 (1997)
7. Braden, R., Clark, D., Shenker, S.: Integrated services in the internet architecture: an overview, RFC 1633 (1994)
8. Carzaniga, A.: Architectures for an Event Notification Service Scalable to Wide-area Networks. PhD thesis, Politecnico di Milano (1998)
9. Eugster, P.T., Guerraoui, R., Sventek, J.: Distributed asynchronous collections: Abstractions for publish/Subscribe interaction. In: Bertino, E. (ed.) ECOOP 2000. LNCS, vol. 1850, pp. 252–276. Springer, Heidelberg (2000)
10. Wroclawski, J.: The use of RSVP with IETF integrated services, RFC 2210 (1997)
11. Joshi, R., Castellote, G.-P.: A Comparison and Mapping of Data Distribution Service and High-Level Architecture (2006),
    http://www.rti.com/docs/Comparison-Mapping-DDS-HLA.pdf
12. Data Distribution Service for Real-time Systems Version 1.2 (2007),
    http://www.omg.org/cgi-bin/doc?formal/07-01-01
13. Berndtsson, M., Hansson, J.: Workshop Report: The First International Workshop on Active and Real-Time Database Systems. ACM SIGMOD Record 25(1), 64–66 (1996)
14. Adelberg, B., Kao, B., et al.: Overview of the Stanford Real-time Information Processor STRIP. ACM SIGMOD Record 25(1), 34–37 (1996)
15. Ramamritham, K., Shen, C., et al.: Using Windows NT for Real-Time Applications: Experimental Observations and Recommendations. In: Proceedings of the Fourth RTAS, Denver, Colombia, pp. 102–111 (1998)
16. Huang, J., Stankovic, J., Towesly, D., Ramamritham, K.: Experimental Evaluation of Real-Time Transaction Processing. In: Proceedings of the 10th RTSS, pp. 144–153 (1989)
17. Shen, C., Gonzalez, O., Mizunuma, I.: User Level Scheduling of Communicating Real-Time Tasks. In: Proceedings of the Fifth RTAS, Vancouver, Canada, pp. 164–175 (1999)

18. Wei, L., Qiang, W., Hongan, W., Guozhong, D.: Adaptive Real-Time Publish-Subscribe Messaging for Distributed Monitoring Systems. Chinese of Journal Electronics, 569–574 (2005)
19. Hoffert, J., Schmidt, D., Gokhale, A.: A QoS Policy Configuration Modeling Language for Publish/Subscribe Middleware Platforms. In: DEBS 2007, Toronto, Canada, pp. 140–145 (2007)
20. Hauer, J.-H., Handziski, V., K?opke, A., Willig, A., Wolisz, A.: A Component Framework for Content-based Publish/Subscribe in Sensor Networks. In: Verdone, R. (ed.) EWSN 2008. LNCS, vol. 4913, pp. 369–385. Springer, Heidelberg (2008)
21. Liu, W., Qiao, Y.: A Visual Specification Tool for Event-Condition-Action Rules Supporting Web-based Environment. In: Proceedings of ICEIS, pp. 246–251 (2008)
22. Araujo, F., Rodrigues, L.: On QoS-Aware Publish-Subscribe. In: Proceedings of DEBS 2002, Vienna, Austria, pp. 511–515 (2002)
23. Corsaro, A., Querzoni, L., Scipiont, S., Piergiovanni, S.T., Virgillito, A.: Quality of Service in Publish/Subscribe Middleware. In: Global Data Management, pp. 1–19. IOS Press, Amsterdam (2006)
24. Eugster, P.T.H., Felber, P.A., Guerraoui, R., Kermarrec, A.-M.: The Many Faces of Publish/Subscribe. ACM Computing Surveys 35(2), 114–131 (2003)
25. Sharifi, M., Taleghan, M.A., Taherkordi, A.: A publish-subscribe middleware for real-time wireless sensor networks. In: Alexandrov, V.N., van Albada, G.D., Sloot, P.M.A., Dongarra, J. (eds.) ICCS 2006. LNCS, vol. 3991, pp. 981–984. Springer, Heidelberg (2006)
26. Gehani, N.H., Jagadish, H.V., Shmueli, O.: Composite event specification in active databases: Model and implementation. In: Proceedings of the 18th VLDB, pp. 327–338 (1992)
27. Demers, A., Gehrke, J., Hong, M., Riedewald, M., White, W.: Towards expressive publish/Subscribe systems. In: Ioannidis, Y., Scholl, M.H., Schmidt, J.W., Matthes, F., Hatzopoulos, M., Böhm, K., Kemper, A., Grust, T., Böhm, C. (eds.) EDBT 2006. LNCS, vol. 3896, pp. 627–644. Springer, Heidelberg (2006)
28. Diao, Y., Altinel, M., Zhang, H., Franklin, M.J., Fischer, P.M.: Path sharing and predi-cate evaluation for high-performance XML filtering. TODS 28(4), 467–516 (2003)
29. Ryll, M., Ratchev, S.: Towards A Publish/Subscribe Control Architecture for Precision Assembly with the Data Distribution Service. In: IFIP International Federation for Information Processing, vol. 260, pp. 359–369, Springer, Boston (2008)
30. Bry, F., Eckert, M., Patranjan, P.-L., Romanenko, I.: Realizing Business Processes with ECA Rules: Benefits, Challenges, Limits. In: Alferes, J.J., Bailey, J., May, W., Schwertel, U. (eds.) PPSWR 2006. LNCS, vol. 4187, pp. 48–62. Springer, Heidelberg (2006)

A Near Real-Time Reporting System for Enterprises Using JavaScript Instrumentation with Inter-colo Event Replication

Timothy Tully

Yahoo!, 701 First Ave., Sunnyvale, Ca 94089, USA
timt@yahoo-inc.com

Abstract. Yahoo! is on track to realize its goal of real-time enterprise-level reporting. Accessing real-time reports allows executives and decision makers to program content and advertising in a way that benefits both the business and the end user. This paper describes our legacy architecture, as well as a new, low latency pipeline. In particular, we show that by using a combination of novel JavaScript instrumentation techniques, as well as an automated, standardized reporting system on top of a near real-time inter-colo event collection mechanism, Yahoo! is nearing its real-time reporting goals.

Keywords: Business Intelligence over Streaming Data, Data Capture in Real Time, Visualization.

Submission Category: Industry Track

1 Introduction

Cost, complexity, and time delays when deploying real-time business intelligence solutions slow down a company's response time and hinders flexibility. On the business side, executives want to speed up data analysis and assure availability of data they value for making business decisions. Organizations which have heavily invested in high latency reporting systems are often reluctant to discard them for new solutions.

Typically, real-time processing systems needed to be sleek and streamlined to keep the number of bottlenecks and points of failure to a minimum. Processing huge amounts of enterprise-scale data was not possible. However, we will describe an architecture that allows us to process data from thousands of web servers worldwide, without making sacrifices in latency.

To that end, near real-time reporting had not been in place at Yahoo! for the first ten years of the company's existence. Web server[1] data collection for the purposes of analytics was, and in some minor instances still is, a difficult problem to solve.

---

[1] Yahoo! uses a specialized version of Apache 1.3[1]. Yahoo! has altered it slightly for performance improvements and to enhance its logging capabilities.

M. Castellanos, U. Dayal, and T. Sellis (Eds.): BIRTE 2008, LNBIP 27, pp. 45–60, 2009.

Consider the amount of data that Yahoo! has to collect on a daily basis. We have thousands of web servers to collect log data from, and many of these machines host multiple properties as well as the international versions of those properties. Moreover, we are collecting data for hundreds of millions of users, spanning 5,000,000 log files each day. The data comes from colos (colocated data centers) around the world, totaling a volume of 10TB per day of data, which is stored into one of our 300TB data warehouses[2].

We quickly realized that we would never be able to have any real-time component to our business reporting unless portions of our architecture changed. This paper will describe three components that have helped us to accomplish real-time reporting:

- Improved instrumentation techniques
- Data collection via colo-to-colo event replication
- Automated reporting of real-time events

In all, when one considers the sheer volume of data, the number of machines and colos we host in, the number of users per day, and the number of event records[3] per user per day that Yahoo! has on a daily basis, extraction of any analytical data in a reasonable amount of time is quite a formidable problem.

1.1 Past BI Pipeline Architecture

By describing our past BI (Business Intelligence) pipeline, shown in Figure 1, the architectural benefits of the new design we describe later are more apparent. Collecting data from thousands of web servers was a massive daily effort. Server log files were cut nightly and copied from each of the colos to a main processing colo, then centrally stored on a number of large-capacity NAS (Network Attached Storage) devices.

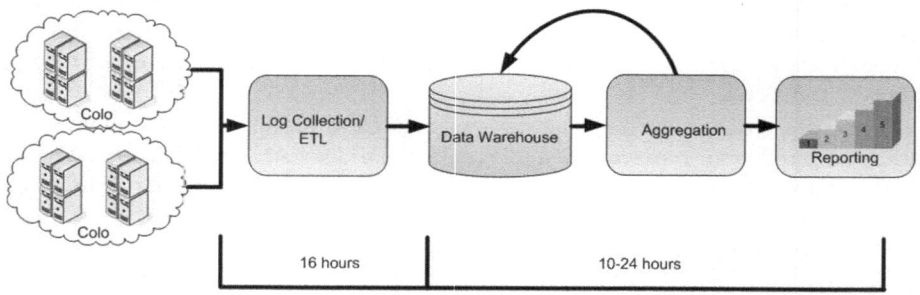

Fig. 1. Legacy Data Pipeline

---

[2] We have other larger petabyte warehouses. [2]
[3] An event to Yahoo! is a pageview and its associated page context (ads, links in the page), or a click (ad or link) and its context.

From there, ETL processing took over and picked up those files for data cleansing and normalization, query term canonicalization and more. We also created pseudo-indices on the data by doing simple lexicographical sorts on columns that had frequent joins on them, via massively parallel and distributed sorting on the log data. The data and each of its respective indexed versions were stored in a 300 TB data warehouse as compressed text files. The time from log collection to ETL to warehouse loading was 16 hours, meaning that post-ETL fact data generated on a Monday would not be available to consumers until 4 p.m. on Tuesday.

The next step in the pipeline was aggregation. Our aggregations were huge, complex parallel processing applications that read data from the data warehouse and wrote aggregated data back to it. This work required enormous numbers of servers to read the daily volume of data. The processing was always network bound, because our data warehouse read text files over NFS, meaning that adding more hardware to the problem would not necessarily fix it. From time to time, we would observe negative scaling because we were network bound, causing our NAS boxes to begin dropping NFS connections and thus invalidating file handles for clients due to contentious reads.

The total time to read data from the warehouse, aggregate it, and load the aggregated data back into the warehouse for digestion by MicroStrategy or other reporting systems was anywhere from 10 to 24 hours. This meant that for some properties, data generated on Monday wouldn't be ready for stakeholders until 4pm on Wednesday.

2   Past Instrumentation Techniques

The instrumentation of web pages was also difficult. Yahoo! had systems that used Apache modules to write custom log data to disk via PHP and Perl APIs. These APIs would allow property developers to collect data such as page context, link view data, and click data. Consequentially, the APIs would also rewrite links in web pages by appending encoded link metadata to them. This created longer URI's than were necessary, making copied and pasted links unnecessarily long and cryptic to read. For example, the following Yahoo! Real Estate URI is 70 characters long, but only 35 characters are the actual URI:

http://realestate.yahoo.com/Sellyourhome;_ylt=An2NGXCedalneQXdiUrrWjykF7kF.

Everything after the semi-colon is used to track the link.

Developers often spent large amounts of time deploying this instrumentation code, as it was a very manual process, often causing developers to wrap each web page link with the API function calls to obtain the rewritten link. Debugging a page's instrumentation was difficult because it often involved manual inspection of raw logs. Also, we had strict ETL filters that periodically made it hard for engineers to understand why certain page views did not appear in the post-ETL data. [4] End-to-end instrumentation of the pages, including coding time, debugging, and validation often took up to two weeks of engineering time.

---

[4] We filter out internal IP traffic, known robot IP blocks, and anomalous records.

The steps of training developers on data systems, instrumenting a property, setting up its ETL processing, writing aggregations and reporting systems together often took a team of five more than three months. In addition, we were faced with reports that were stale, frequently having lead times over of two days after the events occurred.

Yahoo! has many properties in its network, and when a new property came online, we often needed to create an entire reporting team per site in order to report data by product launch. We also realized that real-time reporting was a major business need, and it became necessary to create a simpler methodology for instrumenting, collecting, and reporting our data faster than before.

In the following section, we will present a JavaScript-based instrumentation that tracks user activity in a simpler, more elegant way than server-side tracking. Moreover, we will describe how our instrumentation methodology pipes data into an event replication system that funnels user-event data streams into a single colo, where our real-time reporting application digests event data. The combination of these three components was responsible for bringing our business intelligence reporting down from a two day latency to six minute latency.

3   New Architecture

Investing large amounts of time to instrument properties in the Yahoo! network was not scalable, nor the best use of a property engineer's time. Because of this, we needed to conceive of a new way to track user behavior - in particular, pageviews, link views, and link clicks. It needed to be fast to deploy and relatively unobtrusive, meaning that changes in content shouldn't heavily affect the instrumentation. We also needed a way to get to the data we were collecting much faster than we had in the past.

For the instrumentation piece, we finalized on a lightweight JavaScript library that tracked pageviews, as well as parsed the webpage Document Object Model (DOM)[3] to extract and record link information. This library also tracked subsequent link clicks in the page.

When executed, the instrumentation code in a web page that leveraged this library made HTTP [4] GET requests[5] for images on Yahoo! servers that sent back 1x1 pixel clear JPG files. This is often referred to as beaconing, or inserting web bugs [6].

The images never actually appear in the web page, or the DOM for that matter. We simply want to make requests for images at a particular URI (containing page information), which in effect sends out data to remote servers. The servers that handle the image requests parse the URI's and log them, effectively becoming fact data. Images were used since it allowed us to take advantage of the native JavaScript Image object, which fetches remote images via URIs set in the "src" member.

---

[5] In extreme cases, these URIs would be thousands of bytes long, particularly when there are many link views per page. However, Internet Explorer sets an upper bound [5] of 2k for URIs, meaning that we always had to stay under 2k.

The format of the URI's was set by the library to indicate what user activity - page views, link views, or clicks was occurring. Figure 2 shows a simple example of a page view beacon URI from a Yahoo! UK Movies page:

```
http://eu.rpd.yahoo.com/p?t=1217134478&k=pn%03Now%20Showing&s=97198815&_
r=uk.movies.yahoo.com&u=uk.movies.yahoo.com%2Fnow-showing.html
```

Fig. 2. Example of a UK Movies Beacon

Whenever these image requests were made, associated cookies were sent along in the HTTP header. In the page view case, the URI's contained information about the context of the page, such as background color or page title. In addition, this beacon also contained information about which links were viewed as part of a module. A module can be thought of as a logical section within a web page. Examples of modules can be seen in Figure 3. This example has "World" and "U.S. News" modules, each with five links in them.

We also tracked the position of the link within a section (module), in addition to the typical CTR tracking (link clicks divided by link views). Again, using Figure 3, the link "Nepal's lawmakers abolish the country's monarchy" would have position two in the "World" module. Tracking module usage is important because we're not only interested in what links were viewed, but also which modules were viewed.

The advantage for Yahoo! in this new beaconing system is not that it can report views and clicks - many vendors [7], [8] have been providing this for several years now, often for zero cost. These vendors employ exactly the same methodology described above - JavaScript executing in client browsers, asynchronously sending user behavior data to their servers via HTTP requests for images. The advantage for Yahoo! is the ease and speed by which we can deploy the

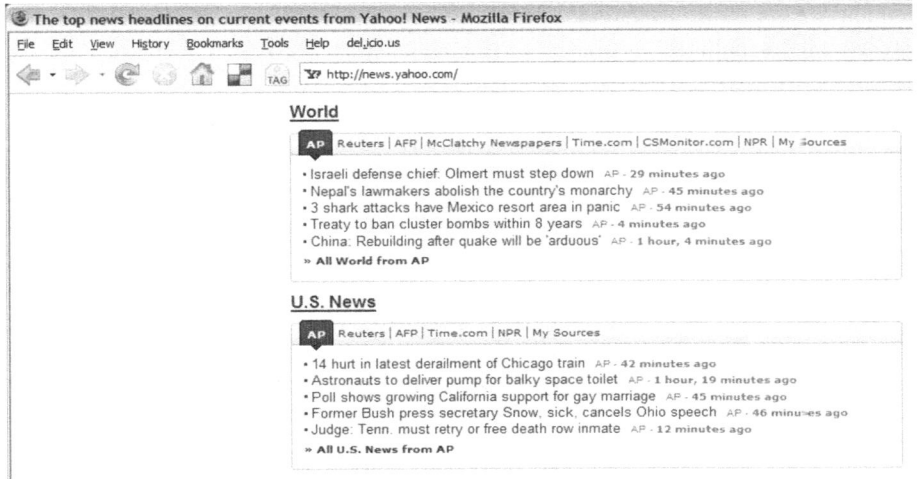

Fig. 3. Example of Modules (Sections) and their Respective Links

instrumentation, and that we can track link view data. The faster we can instrument a page, the faster we can collect usage information about it. Since beacon data goes to a small pool of servers, the new instrumentation system allowed us to collect data from fewer machines. This was in contrast to the legacy system where individual hourly log files were collected from thousands of servers in many colos, which was a huge part of our end to end latency problems. Furthermore, our processing had too many log files to process from too many machines, with too much disk I/O to our NAS devices slowing down the processing.

Not only could the new instrumentation be deployed very quickly, but it could also be deployed by anybody, including non-engineers, because the JavaScript required is simple to write. We created a webpage form-based code generator to write the JavaScript for the person deploying the instrumentation, which we will describe in Section 3.4.

3.1   Instrumentation

Topics surrounding business intelligence in real-time should not focus solely on collecting and reporting data quickly. To access reports and derive insights, data must first be generated. Latency is not just the time it takes to collect and use data, but also part of an end-to-end pipeline. Generating the fact data is the first step in that pipeline. Unless data collection mechanisms can be deployed quickly, the rest of the pipeline will be blocked, waiting for sites to go live, and stakeholders will wait longer to react to business changes.

Our new data instrumentation library is a JavaScript client-side API that provides user behavior tracking capabilities. Using the library, one indicates which modules in a page are to be tracked. The library will then jump to the logical location of those modules by leveraging the browser's DOM API. Then, for each module, it will look for all links that belong to the module. That is, for each module that is to be tracked, the instrumentation will look for all anchor elements (<a>) that are children and descendants of the indicated modules. The first step is to include the library source as a script element:

```
<script src="http://www.yahoo.com/instrument.js"></script>
```

Inside of this file, there is a declared function that returns an object. This function expects a JavaScript object literal, referred to here as the *config* object, to be passed in.[6] Using the contents of this *config* object, the library will automatically begin tracking link clicks on all anchor tags that are children and descendants of the modules to track, as specified by the *config* object. For example, consider the following fragment of a web page in Figure 4: In this example, we have a module that is logically represented by a <div> element. This <div> has two children, which are anchor tags we want to track. Using the library, one would only need to pass the element id of the section we want to track, in this case 'foo'. The library will then beacon data to our web servers, and in the

---

[6] JavaScript doesn't really have proper constructors - objects are byproducts of constructors, which are functions that create objects. [9]

```
<div id="foo">
  <a href="http://www.yahoo.com">Yahoo</a>
  <a href="http://www.facebook.com">Facebook</a>
</div>
```

Fig. 4. Example HTML Source

beacon will be a page view indicator and the links that were seen (in this case Yahoo and Facebook). Furthermore, we will begin tracking link clicks on those two links; each subsequent link click will send a link click beacon.

3.2   Implementation of Library

The instrumentation code placed in a page will operate on the markup of the page based on the DOM tree of that page. We need to write a small piece of JavaScript indicating which node (and corresponding subtree containing links) that we want to instrument. Figure 5 shows an example piece of JavaScript to track the HTML source shown above in Figure 4:

```
<script>
    var keys = {pn:'my page name', bg:'red'};
    var conf = {tracked_mods:['foo'], keys:keys};
    var ins = new YAHOO.i13n.Track(conf);
    ins.init();
</script>
```

Fig. 5. Example JavaScript Instrumentation Code

A line-by-line explanation of this code is shown below:

1. In line 1, we set the page context object. This is a JavaScript object literal, or a JSON[10] object. The keys in this object can be anything, as we'll be encoding it in the beacon format.
2. In line 2, we setup a configuration object, which will be passed to the function that constructs our instrumentation object. The first parameter, *tracked_mods* is an array containing strings that map to element id's in the DOM that are the modules we wish to track. The second parameter, *keys* is a pointer to the object we created in line 1.
3. In line 3, we instantiate an instrumentation object by passing the *conf* object we configured in line 2.
4. In line 4, we call the *init* method on the object we just created, and this will execute the instrumentation logic.

After the init() method is called, the library will examine the page DOM, look at all anchor tags that are child nodes of the element ids indicated in *tracked_mods*, and create a URI formatted to contain data about all links and the context of the page that was passed in.

Furthermore, the API will attach *mouseclick* [11] event listeners to the node elements of *tracked_mods*, and in those listeners, beacon information about the link click. Although we want to track clicks for each link that is part of a module, we only need one event handler per module. This is achieved through the use of event delegation[12]. Using event delegation, the browser allows us to capture mouseclick events during the bubbling phase[13] of event processing. We can then inspect the MouseEvent object for its target, which tells us precisely which link was clicked on.

3.3   Implicitly Collected Data

The API also silently collects many implicit pieces of data. For example, screen resolution and the (X,Y) screen coordinates of link clicks are also captured and passed along in the link click beacons. Those two pieces of data can be used to construct heat maps which overlay mouse click coordinates on top of a web page to indicate user click locations, as shown in Figure 12.

We were also interested in finding further novel ways to measure robot traffic on our sites. One way to do this is to track mouse movements in the browser - something robot traffic cannot generate. We would test pages for mouse movement, and then beacon that data to our servers. This measurement is not done on each page load, but periodically per user.

3.4   Instrumentation Generation

As described earlier, we had a problem with generating instrumentation code quickly because it involved modification of server-side code in PHP. We also needed instrumentation for pages that have short lifetimes, and did not want to invest significant amounts of time to instrument them. To handle this, we created an instrumentation generator that allows developers to fill out web-based forms that automatically generate the JavaScript necessary to instrument a particular page. Figure 6 shows an example of the instrumentation generator. In the background of the page is a form that the developer fills out with details about the page and what needs to be tracked. After submitting the form, a modal window will appear in the browser, containing the code used to instrument the site. The developer or product manager can then simply copy and paste the code.

3.5   Validation of Instrumentation

At this point, we have instrumented our page via JavaScript, and the instrumentation data is asynchronously sent from the browser via HTTP GET requests. This is convenient, since we can monitor outbound traffic from the browser via a browser plugin, very similar to what livehttpheaders[14] does.

For our plugin, we built a Firefox sidebar that inspected traffic via the nsI-HttpChannel interface[15]. We need only to inspect all HTTP headers[16] that leave the browser, and match the destination hosts in the headers with our known beacon server hostnames. If one of the HTTP request hostnames matches, then it is a beacon. We can then decompose the outbound URI and present its contents

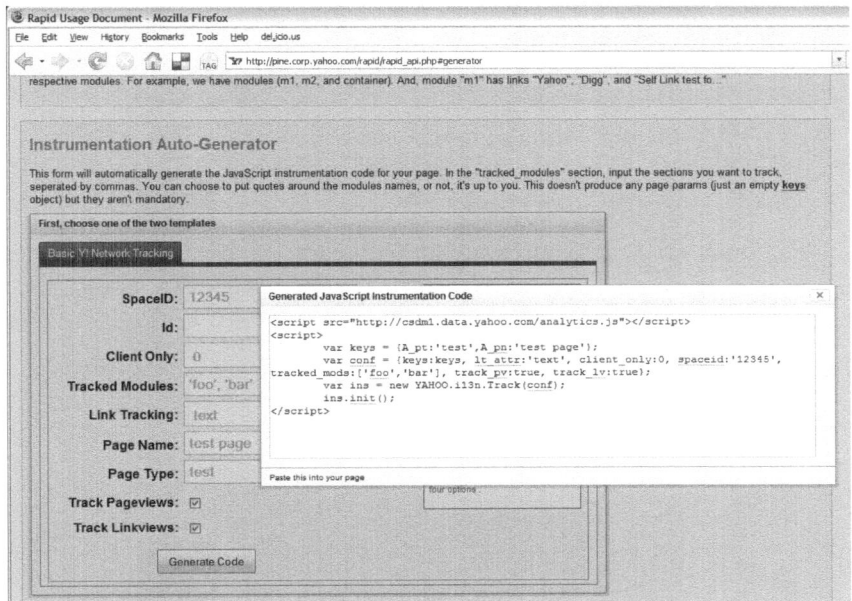

Fig. 6. Instrumentation Generator

in the sidebar. The developer can see in real-time what data is being tracked as it occurs, as well as what the resultant instrumentation on that page will be. An example of the sidebar can be seen in Figure 7.

Each beacon is presented in a tree structure, and each of the elements of the beacon becomes a node in the tree. Some nodes have their own subtrees, such as the module/linkview data. Each logical module has its own node in the tree, and each link that is a member of that module is shown as a child of the module. For example, the module named "Reviews" has a link to "profiles.yahoo.com...".

The significance of being able to view live instrumentation as it happens cannot be understated. In the past, we had to inspect the data as it was generated on the web server by parsing the server logs and manually reviewing those files. Having a tool in the browser that reveals what we are instrumenting as it happens results in large time savings for the developer and results in faster deployment of pages. Also, it assures the developer that the code placed in the page is indeed functioning and is sending out the exact data that we want to track.

3.6   Data Collection

In this section we will describe Yahoo!'s method to provide downstream consumers with access to beacon web server log fact data in near real-time (seconds in colo) and low-latency (sub-six minute), using an inter-colo event replication system.

Consider the processing of data in any one of the colos that Yahoo! uses worldwide. It can have thousands of machines that have web server logs for collection and subsequent ETL and aggregation.

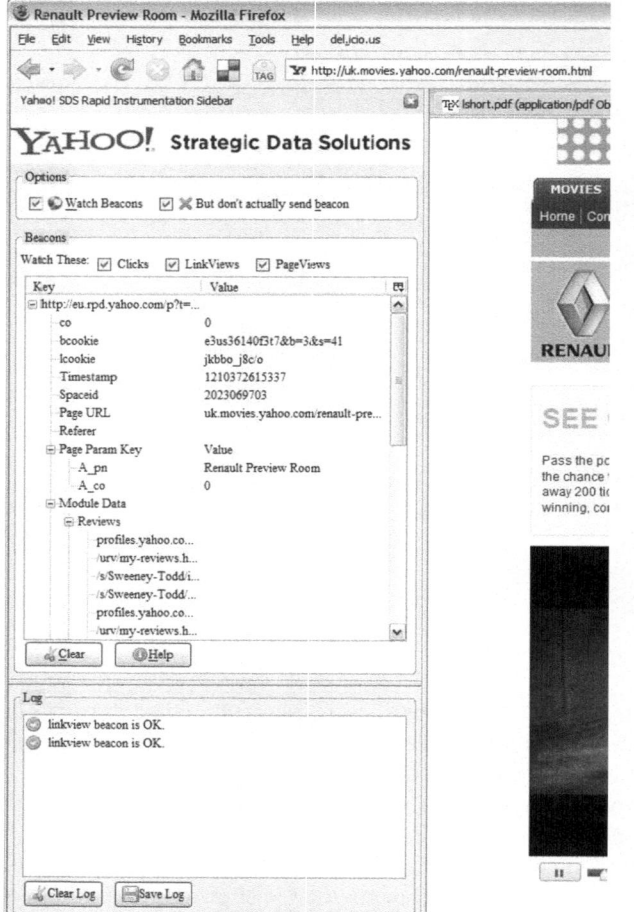

Fig. 7. Instrumentation Firefox Sidebar

The standard implementation for retrieving data on each of those machines would be to write software that cuts log files periodically, batches them, and copies them over to one colo for ETL, aggregation, and reporting. This is operationally difficult to maintain since we would have to run this process on every single web server, as well as have a distributed processing system to coordinate the file copies. Also, it would prevent us from meeting any real-time data access requirements in a scalable way.

Another way to attack this problem would be to bypass ETL altogether, and aggregate data on each of the web servers and then merge that downstream. However, this would add extra load (to compute aggregates inline) on the servers, not to mention the fact that we still need the raw fact data for downstream consumers.

However, if we were able to process the log data on a smaller, separate cluster of machines, processing would be more efficient, less load would be placed on the

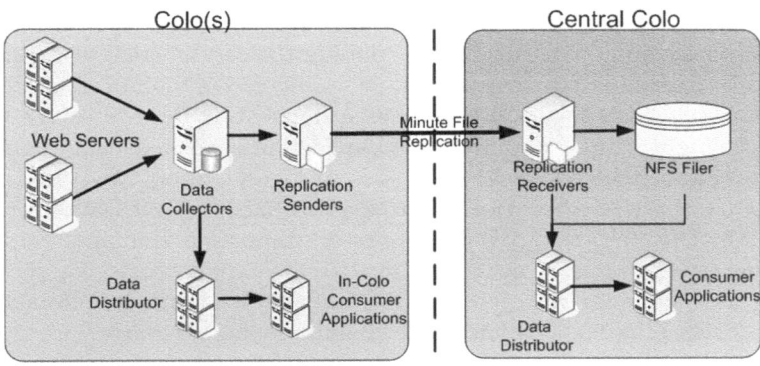

Fig. 8. Current Log Collection Architecture

web servers, and data processing could be centralized to a smaller set of nodes. The problem with this solution is getting the upstream fact data, in this case raw server log files, to that central processing pool.

The architecture diagram in Figure 8 illustrates how we solved our low-latency log data collection problems. In the diagram we show a typical Yahoo! colo on the left, seperated by a dashed line which represents the logical separation between web server colos from our central data-collection colo. Yahoo! has many colos around the world, just like the one on the left side of the dashed line. Data from each of the colos is moved to our central processing colo on the right.

Typical Yahoo! colos contain hundreds and even thousands of web servers. Each of these Apache web servers is fitted with an Apache module [17] whose responsibility is to stream live events (HTTP requests) off of the server, and onto a Data Collector (DC). The Data Collector batches data and forwards it onto a Replication Sender, whose responsibility is to forward data (events) out of the colo and into our main collection colo.

3.7 In-Colo Collection of Real-Time Data

Note that data can be streamed to applications either in the same colo where it is generated, or in a centralized location, where data from all colos is merged into one stream. In the in-colo scenerio, the Data Collector has the capability to stream data to another set of boxes, namely the Data Distributor. The Data Distributor is another machine whose responsibility is to replicate the events to any number of consumers who want to read live data. Typically, it takes approximately six seconds for events to reach the consumer once Apache receives them. Data consumed by in-colo applications comes in as per-event records, streaming over shared TCP socket resources. The downside to this solution is that this six-second data will be colo-only, and we would not have access to worldwide colo data.

3.8   Centralized Collection of Real-Time Data

For those whose don't want to consume data in the colo where it was generated, data can be fetched at the central colo, providing access to data from all worldwide colos. Revisiting Figure 8, we can see a Replication Receiver in the Central Colo, receiving streaming data from the Replication Senders in each of our colos. The Replication Receiver's sole purpose is to digest the events, store them on a temporary space (NAS device), and pass the data on to a Data Distributor cluster. The Data Distributor's role in the central colo is similar to its role for the in-colo setup, that is, to distribute data to client applications. The advantage to doing ETL and data aggregation in the central colo is that data from all colos can be processed in one location in the six minute latency range.

3.9   Data Filtering

Receiving streaming data from the Data Distributor in the central colo would allow a consumer to process all events from all web servers in all of the colos, as a merged stream of data. However, it is rare that we need to see data for all Yahoo! properties. Often, we're interested in a handful of properties or fewer. To accomplish this, a developer may specify an event filter on the Data Distributor. For example, the consumer may want to only process events from specific servers or events that have URI's matching some specified regular expression.

Furthermore, developers may not want all columns that compose a record; perhaps only one or two columns out of twenty are necessary. In the same way we specified event filters on the Data Distributor, we can do the same thing with a projection specification. Using configuration files, we can tell the Data Distributor exactly which columns of a record we want to digest.

The final high-level architecture for our real-time data collection and reporting can be seen in Figure 9.

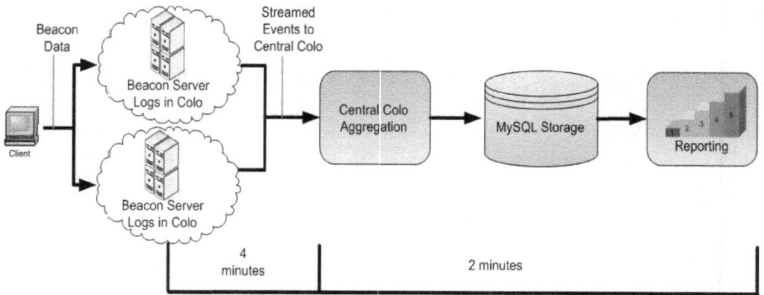

Fig. 9. Real-Time Data Pipeline

3.10   Reporting and Visualization

Real-time data is useless without fast aggregation and insightful data reporting. In this section we'll describe a methodology for aggregating fact data on the fly and reporting it in a web browser using an automated data refresh without page reloading.

Much research has been devoted to the study of streamed data management and processing [18], [19]. In our architecture, we have developed a real-time reporting system that reads streaming data from a low latency pipeline in a centralized colo scenario. Similar to existing stream data processing architectures such as STREAM[20], we have developed continuous-processing data aggregation clients that read from the streaming event pipe, bypassing ETL. These processes only store aggregate data, and fact data is then discarded.

Any number of these clients can run in parallel as part of a larger cluster, consuming data off of the Data Distributor, as shown in Figure 8. Since the Data Distributor is multiplexing the fact data across the clients in the cluster, each fact record is a discrete event and guaranteed to be sent to only one client in the pool. This is done to avoid over-counting in the aggregation process.

Each of the clients in the cluster aggregates data as it comes in, using a static list of metrics, such as page views, link and ad clicks, unique users, and more. Each of these aggregates is segmented by page within the Yahoo! network. We have metrics for each property, even within an individual page in a property. For example, we would have page view and unique user counts for both the "Inbox" and "Search" pages in Yahoo! Mail. Each of those pages would have their own unique page identifier, which is just a string uniquely identifying those pages.

The aggregate data is then stored in a clustered MySQL database every minute, segmented by page identifier. On top of the MySQL cluster is a web service layer that returns JSON data structures with the metrics needed for the reporting page. An example of the resultant real-time reporting page is shown in Figure 10.

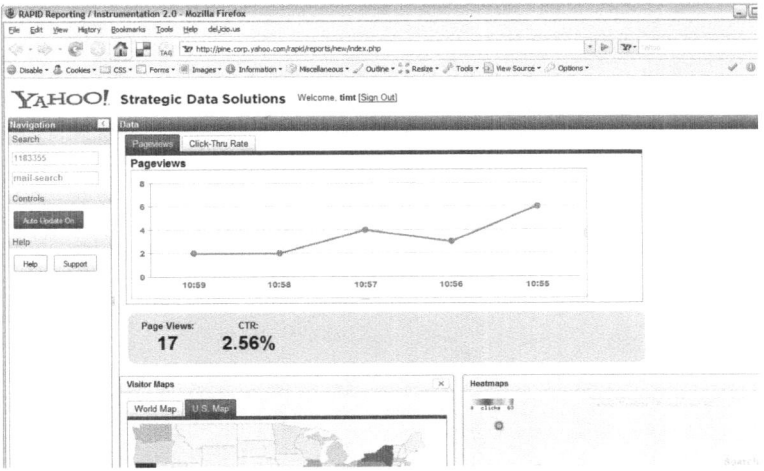

Fig. 10. Pageview Counts Over Time in a Near Real-Time Reporting Portal with Continuous Updating

3.11   Example Use Case

One strong example of leveraging the real-time reporting system is for programming news content and advertising on Yahoo! News. Editors for that page need to know in real-time which stories are performing (based on link click-through rate to the stories). Using the reporting pages shown above, the editors can

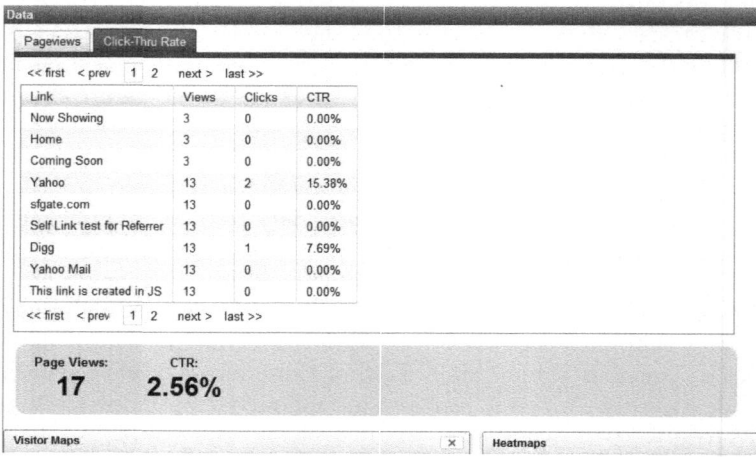

Fig. 11. Link Performance in a Near Real-Time Reporting Portal with Continuous Updating

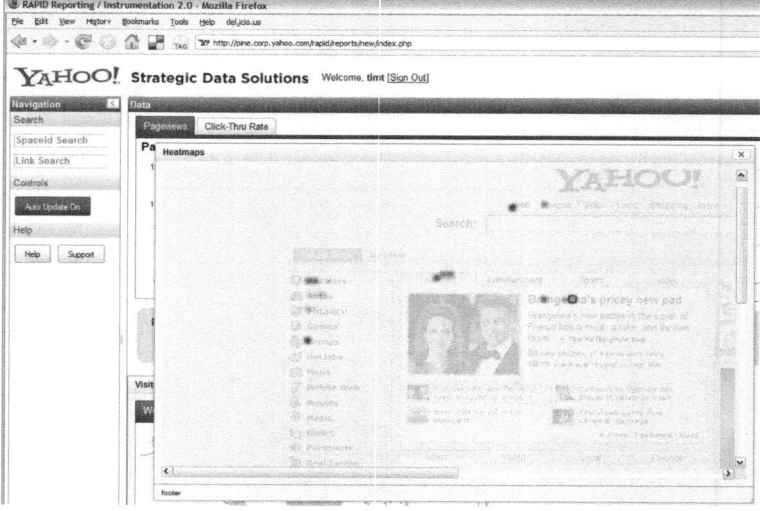

Fig. 12. Mouse Click Heat Maps in a Near Real-Time Reporting Portal with Continuous Updating

quickly see which links are popular, and add and remove content in real-time based on that data.

The reporting page will continually update itself every thirty seconds via AJAX [21] calls that fetch the JSON data from the backend web service. The web service fulfills the web service requests by making queries to the MySQL database.

The report refreshes itself through JavaScript's setInterval function, which makes the web service call. On web service call return, we get a new set of data to update the tables and graphs within the page. The tables and graphs were rendered using the Yahoo User Interface (YUI)[22]. By updating page data via JavaScript, we can refresh the page content without having to reload the entire page. The Adobe Flash line graph shown in Figure 10 will shift to the right and update itself with the most recent data set. The graphs will not flicker or disappear, as they are animated and seemlessly update themselves in a non-disruptive way. We also plot the geographic source of incoming traffic, as shown in the bottom left of Figure 10. Similarly, the link performance table showing linkviews, clicks, and effective click-through rate (CTR), shown in Figure 11, will update itself as well. This functionality allows stake-holders to watch the reporting page report link performance in real-time without refreshing pages.

4   Results and Conclusion

Our data reporting needs have changed, and recently we have moved towards scalable, lower latency systems, in particular near real-time reporting. By changing our methods of instrumentation, data collection, and reporting, Yahoo! has reduced its business intelligence latency from multiple days to a matter of minutes.

In order to make this change in reporting lead time, several innovative, fundamental changes were made to our data pipeline. First, our instrumentation needed to be more robust; continually wrapping web site links with function calls in server side application languages required huge developer time investments, and often led to erroneous results.

To that end, we introduced a feature-rich JavaScript instrumentation framework that allowed for faster, simpler ways to track user behavior and web site metrics, as well as more advanced validation tools. This library has brought instrumentation engineering time down from weeks to minutes.

In addition, our data collection mechanism needed to be transformed from a divide and conquer raw log collection scheme, to a platform that streams events between and within colos. We have shown that this novel system can reduce data collection time from the next day, to minutes.

Finally, we made changes in reporting to introduce real-time aspects to it, as opposed to static pages that require manual page refreshes. This was achieved by creating web pages that seamlessly refresh themselves via asynchronous polling for data updates. In addition, the visualization of our reporting changed; we created graphs and tables that elegantly present the data to reflect its real-time aspects.

Acknowledgments

The author thanks the reviewers of this paper for their invaluable recommendations, in particular James Merino and Amit Rustagi.

References

1. Apache web server software, http://www.apache.org
2. Claburn, T.: Yahoo Claims Record With Petabyte Database, InformationWeek (2008), http://www.informationweek.com/news/software/database/showArticle.jhtml?articleID=207801436
3. Document Object Model, http://www.w3.org/DOM/
4. Hypertext Transfer Protocol - HTTP/1.1, http://www.w3.org/Protocols/rfc2616/rfc2616.html
5. Maximum URL Length in 2,083 characters in Internet Explorer, http://support.microsoft.com/kb/208427
6. Web bug - Wikipedia, http://en.wikipedia.org/wiki/Web_bug
7. Google Analytics, http://www.google.com/analytics/features.html
8. Omniture Web Analytics, http://www.omniture.com/en/products/web_analytics
9. Private Members in JavaScript, http://www.crockford.com/javascript/private.html
10. JSON, http://www.json.org
11. Mouse Events in the browser, http://www.quirksmode.org/js/events_mouse.html
12. Web Browser Event Delegation, http://developer.yahoo.com/yui/examples/event/event-delegation.html
13. Event Bubbling, http://www.quirksmode.org/js/events_order.html
14. Livehttpheaders, http://livehttpheaders.mozdev.org
15. nsIHttpChannel interface API, http://xulplanet.mozdev.org/references/xpcomref/nsIHttpChannel.html
16. Header Field Definitions in Internet RFC 2616, http://www.w3.org/Protocols/rfc2616/rfc2616-sec14.html
17. Apache Module description, http://httpd.apache.org/modules/
18. Cranor, C., Johnson, T., Spatscheck, O., Shkapenyuk, V.: Gigascope: A Stream Database for Network Applications. In: Proceedings of the 2003 ACM SIGMOD International Conference on Management of Data, pp. 647–651. ACM, New York (2003)
19. Zdonik, S.B., Stonebraker, M., Cherniack, M., Çetintemel, U., Balazinska, M., Balakrishnan, H.: The Aurora and Medusa Projects. In: IEEE DE Bulletin, pp. 3–10 (2003)
20. Stanford Stream Data Manager, http://infolab.stanford.edu/stream/
21. AJAX, http://en.wikipedia.org/wiki/AJAX
22. Yahoo User Interface, http://developer.yahoo.com/yui

A Hybrid Row-Column OLTP Database Architecture for Operational Reporting

Jan Schaffner, Anja Bog, Jens Krüger, and Alexander Zeier

Hasso Plattner Institute for IT Systems Engineering,
University of Potsdam, August-Bebel-Str. 88,
D-14482 Potsdam, Germany
{jan.schaffner,anja.bog,jens.krueger,alexander.zeier}@hpi.uni-potsdam.de
http://epic.hpi.uni-potsdam.de

Abstract. Operational reporting differs from informational reporting in that its scope is on day-to-day operations and thus requires data on the detail of individual transactions. It is often not desirable to maintain data on such detailed level in the data warehouse, due to both exploding size of the warehouse and the update frequency required for operational reports. Using an ODS as the source for operational reporting exhibits a similar information latency.

In this paper, we propose an OLTP database architecture that serves the conventional OLTP load out of a row-store database and serves operational reporting queries out of a column-store database which holds the subset of the data in the row store required for operational reports. The column store is updated within the transaction of the row database, hence OLTP changes are directly reflected in operational reports. We also propose the virtual cube as a method for consuming operational reports from a conventional warehousing environment.

The paper presents the results of a project we did with SAP AG. The described solution for operational reporting has been implemented in SAP Business ByDesign and SAP Business Warehouse.

Keywords: Real-time decision support, real-time operational data stores, data warehouse evolution.

1 Introduction

Inmon distinguishes *operational* and *informational* reporting [10]. According to his classification, informational reporting is used to support long-term and strategic decisions, looking at summarized data and long-term horizons. Informational reporting is typically done using a data warehouse. Operational reporting, in contrast, is used to support day-to-day decisions, looking at the data on a more detailed level than in informational reporting and taking up-to-the-minute information into account where possible. When using a data warehouse for operational reporting, the following implications must be considered:

- The data warehouse must be designed to the level of granularity of the operational data, which might not be desirable.

M. Castellanos, U. Dayal, and T. Sellis (Eds.): BIRTE 2008, LNBIP 27, pp. 61–74, 2009.
© Springer-Verlag Berlin Heidelberg 2009

- Updates to the operational data must be replicated into the warehouse on a frequent basis. While ETL approaches exists that support frequent updates of the warehouse (e.g. Microbatch [3]), the query performance of the warehouse suffers dramatically during the frequent updates.

At the same time, operational reporting cannot be done on operational data either: Analytical queries are long-running compared to OLTP (On-Line Transactional Processing) queries, and the locks of the analytical queries might result in a decreasing transaction throughput, since short running OLTP queries have to queue until analytical queries touching the same data release their locks. Hence, a solution is called for that enables operational reporting without affecting neither operations nor standard warehousing for informational reporting.

As the first contribution of this paper, we present a hybrid store OLTP engine that consists of a traditional row-oriented relational database and a column-oriented database which holds replicas of all OLTP data relevant for operational reporting. The column store is updated within the transaction of the relational database, making all updates available for operational reporting immediately. The second contribution is the *virtual cube* construct which is used to map the transactional data model exposed in the column store to an OLAP (On-Line Analytical Processing) structure, providing analytics-style data exploration methods for our operational reporting architecture. The virtual cube is also useful when trying to consolidate operational reporting across multiple transactional systems. However, the virtual cube does only provide a different view on the columnar operational data, while data cleansing and harmonization techniques are not applied.

The remainder of this paper is structured as follows: Section 2 reviews the benefits of column store databases for analytics-style workloads. Section 3 then describes our hybrid architecture consisting of a row-oriented database for conventional OLTP and a column-oriented database for operational reporting. We also describe how we implemented this architecture with a prototype on SAP Business ByDesign, a hosted enterprise resource planning solution for small and medium enterprises. Section 4 outlines the concept of the virtual cube and describes our implementation in SAP Business Warehouse. Section 5 surveys related work, outlines the differences of our hybrid store OLTP engine when compared to operational data stores, and positions the virtual cube construct in the Business Intelligence landscape. Section 6 concludes the paper.

2 Column Databases

In 1985, Copeland and Khoshafian introduced the decomposed storage model (DSM) [8]. Each column is stored separately (see Figure 1) while the logical table schema is preserved by the introduction of surrogate identifiers (which might be implicit).

The use of surrogate identifiers leads to extra storage consumption, which can be overcome, for example, by using the positional information of the attributes in the column as identifier. Encoding approaches exist that avoid the redundant

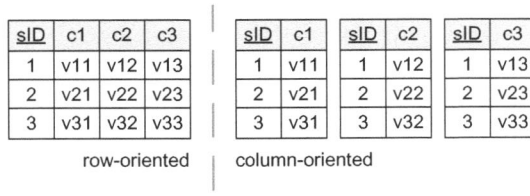

Fig. 1. Row- vs. Column-oriented Storage

storage of attribute values, e.g. null values, or in the case of columns where only a small amount of differing values exists. We will touch briefly on these later in this section. The idea of column store databases has been implemented in multiple projects, for example, MonetDB [4], C-Store [20], or BigTable [6].

Column stores are especially useful for analytical workloads, which is due to the fact that analytical applications are largely attribute-focused rather than entity-focused [1], in the that sense only a small number of columns in a table might be of interest for a particular query. This allows for a model where only the required columns have to be read while the rest of the table can be ignored. This is in contrast to the row-oriented model, where all columns of a table – even those that are not necessary for the result – must be accessed due to their tuple-at-a-time processing paradigm. Reading only the necessary columns exhibits a more cache-friendly I/O pattern, for both on-disk and in-memory column store databases. In the case of an on-disk column store, few disk seeks are needed to locate the page of a block that has the first field of a particular column. From there, data can be read in large blocks. In the case of an in-memory column store, sequentially laid out columns typically ensure a good L2 cache hit ratio, since modern DRAM controllers use pre-fetching mechanisms which exploit spatial locality. Abadi demonstrates that these advantages cannot be achieved when emulating a column-store using a row-based query executor, i.e. using separate tables and indexes for each column in a logical table [1].

For queries with high projectivity, column-oriented databases have to perform a positional join for re-assembling the different field in a row, but projectivity is usually low in analytical workloads. Column stores can leverage the read-only nature of analytical queries by partitioning (i.e. scale-out): accessing single columns for values and computing joins can be massively parallelized when distributing data on multiple machines. Two basic approaches for the distribution of data onto multiple machines exist. These are horizontal and vertical fragmentation, see Figure 2. Horizontal fragmentation separates tables into sets of rows and distributes them on different machines in order to perform computations in parallel. It has been introduced, for example, to solve the problem of handling tables with a great number of rows, like fact tables in a data warehouse [16]. Column-orientation facilitates vertical fragmentation, where columns of tables are distributed on multiple machines. SAP BI Accelerator (BIA), the column store component of our hybrid architecture, uses both techniques [13].

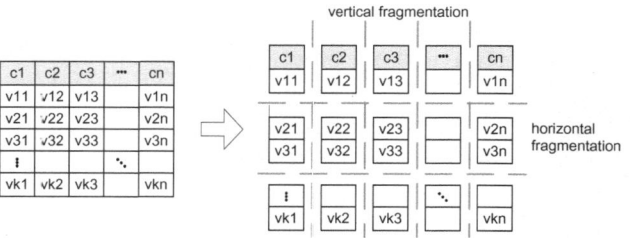

Fig. 2. Horizontal and Vertical Fragmentation

In column stores, compression is used for two reasons: saving space and increasing performance. Compressed columns obviously consume less space on disk or in-memory. Performance can be increased by making the colum-oriented query executor aware of the type of compression that is being using. Decompressing when reading data, however, puts more load on the CPU. The main trade-off in compression is compression ratio vs. the cost for de-compression. However, a widening gap between the growth rate of CPU speed and memory access speed can be observed [14]. While CPU speed grows at a rate of 60 percent each year, the access time to memory (DRAM) increases less than 10 percent per year. This trend argues for the usage of compression techniques requiring higher effort for de-compression, since CPU resources grow and the cost for I/O is high. However, when taking into account that aggregate functions are typical operations in analytical queries, the use of compression techniques on top of which these functions can directly be performed without de-compressing becomes more appealing. Abadi et al. [2] have characterized and evaluated a set of compression techniques working particularly well under this assumption, e.g. run-length encoding (RLE) or bit-vector encoding.

Data compression techniques exploit redundancy within data and knowledge about the data domain for optimal results. Compression applies particularly well to columnar storage, since all data within a column a) has the same data type and b) typically has similar semantics and thus a low information entropy (i.e. there are few distinct values in many cases). In RLE the repetition of values is compressed to a (value, run-length) pair. For example the sequence "aaaa" is compressed to "a[4]". This approach is especially suited for sorted columns with little variance in the attribute values. For the latter if no sorting is to be applied, bit-vector encoding is well suited. Many different variants of bit-vector encoding exist. Essentially, a frequently appearing attribute value within a column is associated with a bit-string, where the bits reference the position within the column and only those bits with the attribute value occurring at their position are set. The column is then stored without the attribute value and can be reconstructed in combination with the bit-vector. Approaches that have been used for row-oriented storage are also still applicable for column-oriented storage. One example is dictionary encoding, where frequently appearing patterns are replaced by smaller symbols.

Currently, SAP BIA uses integer and dictionary encoding in combination with bit-vector encoding. Each existing value for an attribute is stored in a dictionary table and mapped to an integer value. Within the columns only the integer values are stored. As a first advantage attribute values existing multiple times within a column reference the same row within the dictionary table. Thereby redundant storage of attribute values is eliminated and only redundancy of the integers referencing the same attribute value occurs. The second advantage is that the integers used for encoding consume less storage space than the actual attribute values.

Due to the compression within the columns, the density of information in relation to the space consumed is increased. As a result more relevant information can be loaded into the cache for processing at a time. Less load actions from memory into cache (or disk into memory) are necessary in comparison to row storage, where even columns of no relevance to the query are loaded into the cache without being used.

3 A Hybrid Architecture for Operational Reporting

In this section, we will describe a hybrid architecture, consisting of both a row store and a column store, for performing operational reporting in an OLTP system. First, we argue the necessity for operational reporting as we have observed it at real companies. We will then describe our architecture showing how it is able to address those needs.

3.1 Necessity of Operational Reporting

Inmon first introduced the notion of operational reporting in 2000 [10], characterizing it as analytical tasks that support day-to-day decisions, looking at the data on a more detailed level than in informational reporting and taking up-to-the-minute information into account where possible. However, operational reporting does not only exist in literature: in the course of our research, we had the opportunity to work with a number of SAP customers and to understand what types of applications they implement both on top of their ERP systems and in their data warehouses. One example are planning activities, for example when trying to forecast sales for products offered. Such planning is important to estimate cash-flow and to drive manufacturing planning in order to cater for the expected demand. This plan will then be tracked against actual sales and will frequently be updated. While it would be conceivable to trickle-load sales order data into a data warehouse every night, we have often seen the requirement to look at the sales data on an intra-day basis to support decisions related to production planning or financial planning.

3.2 Architecture

This section introduces an architecture for operational reporting where no replication into a data warehouse occurs, but data is accessed directly in the transactional systems when queried.

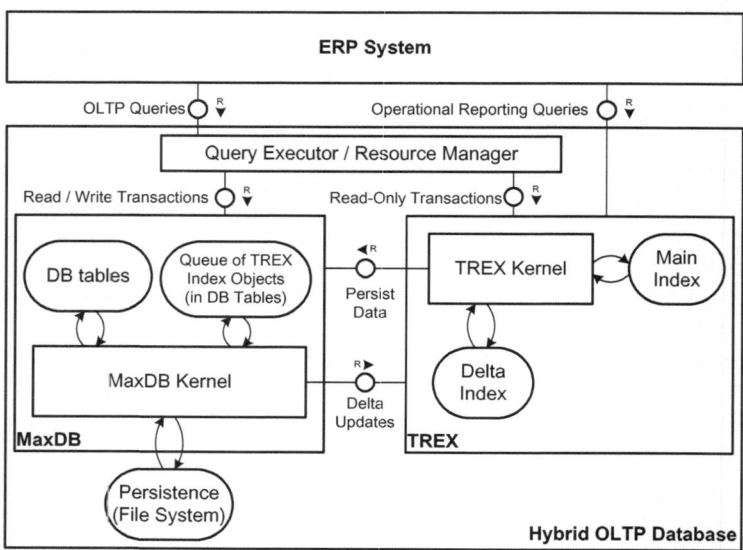

Fig. 3. Integration of TREX with MaxDB

Our architecture consists of both a row-oriented database for the entity-oriented OLTP operations (i.e. "full-row" queries) and a column-oriented database to handle the operational reporting queries. As the row-oriented database we use SAP MaxDB, since it is the database underlying SAP Business ByDesign, the mid-market ERP software we extended in our project. It fully supports ACID[1]. As the column-oriented database we used SAP's Text Retrieval and Information Extraction engine (TREX), which is the engine underlying SAP BIA. TREX has originally been developed as a text search engine for indexing and fast retrieval of unstructured data. In our work, we use it as the main memory column database that answers operational reporting in Business ByDesign. Figure 3 shows how TREX is integrated with MaxDB and how read and write access is distributed between TREX and MaxDB.

Requests that change data or insert new data are handled by MaxDB, which ensures the ACID properties. The MaxDB Kernel stores the changes in the database tables and manages so-called "queue tables" for TREX. Those queue tables contain information about data that is relevant for operational reporting and that has been updated or inserted in MaxDB. TREX is notified about the changes and can update its own data with the help of the queue tables. This happens within the same database transaction using on-update and on-insert triggers, which are created inside MaxDB when the system is configured. The triggers fire stored procedures which forward the queries to TREX. Accordingly, TREX and MaxDB share a consistent view of data.

---

[1] Atomicity, Consistency, Isolation, Durability.

Queries for operational reporting and some read-only transactions go directly to TREX. TREX manages its data in so-called *main indexes*. The main index holds a subset (i.e. the OLTP data that must be available for operational reporting) of the database tables in MaxDB, except that the schema is flattened and the data is stored column-wise. The advantages of columnar data structures have already been discussed in section 2. Since the main index is highly optimized for read access, TREX holds a *delta index* to allow fast data retrieval while concurrently updating its data set. All updates and inserts taken from the queue tables are collected in the delta index. When responding to a query, data in the delta index as well as the main index is accessed and the results are merged together in order to provide a consistent view of the entire data set compared with the database tables of MaxDB. The delta index is not compressed to allow for fast writes, while it must be able to provide a reasonable read performance. In the current implementation delta indexes are organized as a B-tree in memory. It is important that delta indexes do not grow larger than a certain size limit, since the read time from the delta index should not exceed the read times from the main index of a given table (main and delta index are read in parallel). Upon reaching a certain size (e.g. 5% of the size of the main index) or pre-defined intervals the delta index is merged with the main index. To do so, a copy of the index is created in memory along with a new, empty delta index for that copied index. The queries coming in during the merge will be run against this structure. In particular, the new delta index that receives all inserts and updates during the merge. After the merge of the original delta index and the main index, the new delta index becomes the delta index for the merged index. The copy of the original main index in memory is now discarded. This procedure for merging a delta index with a main index ensures that neither read nor write accesses to TREX are blocked during merge time.

In our implementation, we also extended MaxDB to serve as the primary persistence for the TREX indexes (i.e. the tables), as opposed to using the standard file system persistence of TREX. The reason is that in case of a disaster both MaxDB and TREX can be recovered to a valid state using MaxDB's recovery manager. Therefore, the queue tables in MaxDB described above also serve as a recovery log for the TREX delta indexes. A possible direction for future research would be to conceptually move the delta index in the column store completely into the row store, instead of buffering the column store delta index in queue tables in the row store. Fractured Mirrors [17] and C-Store [20] provide a row-oriented delta mechanism, but do not propose ACID transactions as we do in the architecture presented here.

4   Virtual Cube

In the previous section we described our architecture for operational reporting in OLTP systems using both a row-database for high volume OLTP writes and a column store for handling the long running operational reporting queries. In this section, we will describe the second contribution of this paper, a virtual cube which allows for seamless consumption of operational reports from a

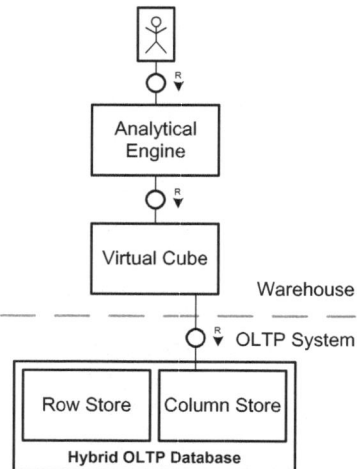

Fig. 4. The Environment of the Virtual Cube

data warehouse environment, in the way that the same interface is provided as for a standard cube. Yet, accessing the operational reports through the data warehouse imposes the restriction of pre-defining the reports in advance. An alternative user interface to the front-end described in this section are tools for data exploration. Such tools allow to navigate through a star schema while operating on large cubes (for e.g. SAP BusinessObjects Polestar which integrates with TREX).

Figure 4 provides a high-level overview of the target architecture. The analytical engine accesses data through the virtual cube. The virtual cube provides the same interface for analysis as standard cubes in a data warehouse. This includes navigation along various levels of hierarchy (i.e. drill-down and roll-up) as well as slicing and dicing along different dimensions. In our prototypical implementation, we created a virtual cube that plugs into the OLAP engine of SAP BI. In consequence, all the reporting front-ends supported by SAP BI can be used to launch queries against the OLTP data. Available front-ends include HTML reports and Microsoft Excel-based reports. In the case of SAP BI, predefined queries must be run inside these reporting front-ends. These queries can be specified graphically (using a query design tool) or using MDX[2].

In comparison to traditional cubes in a warehouse, the virtual cube does not store any data. Instead, it is a collection of functions that are executed during the run-time of a report. The operational reporting queries from the reporting front-end are sent to the OLAP engine, which then executes OLAP queries against the virtual cube. The latter transforms the incoming queries into queries against TREX. This transformation is mainly concerned with passing the requested

---

[2] Multidimensional Expressions:
   http://msdn2.microsoft.com/en-us/library/ms145506.aspx

key figures (i.e. facts that are to be aggregated) and dimensions to TREX. For key figures, the type of the desired aggregation (e.g. sum, average, minimum, maximum) must be passed to TREX along with the name of the attribute, whereas dimensions (e.g. product, region, quarter) are just listed. All dimension attributes are added to the GROUP BY part of the query against TREX. While the strategy for obtaining the best response times for operational reporting is to push as much computation as possible down to TREX, predicate filtering on attributes is handled by the OLAP engine of SAP BI in our current prototypical implementation. While it would technically be possible to push filtering down to the TREX layer, we chose to use the result filtering capabilities of the OLAP engine due to some of the particularities of TREX's query API that would have required to implement a more elaborate query rewriter between the OLAP engine and TREX. In some cases this also results in the necessity for post-aggregation in the OLAP engine after a query against TREX has returned. This is similar to materialized view selection for multi-cube data models [18] where aggregates are retrieved from several cubes before post-aggregation occurs. In our case, however, there are no cubes containing aggregates and the virtual providers queries the OLTP records directly, since the OLTP data in TREX is stored on the highest possible level of granularity (i.e. the data does not contain aggregates but is on item-level). Therefore, aggregations required for different levels within the hierarchy of a dimension are computed on-the-fly using up-to-date values from the OLTP system. Our experiments show that on-line aggregation is feasible for data sets of small- and medium businesses: In case study that we have conducted with a medium-size brewery, the average response time for a query encompassing an aggregation of 10 million rows in a table (grouping by four attributes) was 2.1 seconds for a result set containing 200 tuples on a commodity server. Due to space restrictions we are unable to provide a detailed performance evaluation for online aggregation in this paper.

5   Related Work

In this section, our hybrid architecture will be positioned among the existing data warehousing architectures that we are aware of, especially those having a focus on "real-time" or operational data. We will also address work related to the virtual cube described in the previous section.

5.1   Common Data Warehouse Architectures

The general architecture of data warehouses is well known. The characteristics defined by Inmon, i.e. that data in a warehouse must be subject-oriented, integrated, time variant, and non-volatile [11, p. 31], led to an architecture separating operational and analytical data. Data in OLTP systems is organized according to the relational model (defined by Codd [7]), i.e. data is highly normalized in order to ensure consistency and to run day-to-day operations on these systems. OLAP systems, in contrast, organize data according to the dimensional model,

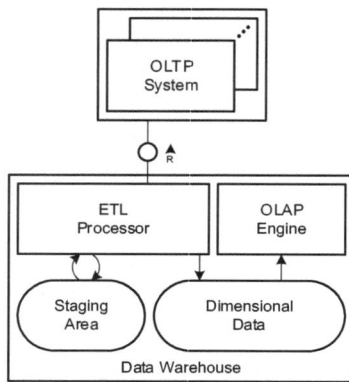

Fig. 5. Traditional Data Warehouse Architecture

using for example the star or snowflake schema. The reason for this is mainly the wish to achieve the best query performance for both OLTP and OLAP.

This leads to an architecture as it is shown in Figure 5. The data warehouse contains an ETL processor which extracts data from various OLTP sources into a staging area, where data transformations for cleansing and integration are applied. Once this process has been completed, the ETL processor stores the data according to a dimensional data storage paradigm, so that an OLAP engine can run queries against this dimensional data store.

With the proliferation of BI technologies, this general architecture has been extended with concepts such as data marts or Operational Data Stores (ODS). Data marts aim at decentralizing warehouses in order to optimize performance around certain subject areas [11]. The downside is that in data mart architectures, the data warehouse cannot provide a consolidated view on all data relevant for strategic decision making in an enterprise, which was the original intent of data warehouses. ODSs store OLTP data, often using an integrated schema, i.e. the ETL steps of data mapping and cleansing are applied before moving data into an ODS. The result is increased timeliness of the data on which reporting can be done, as well as the possibility to work on line-item level in case the ODS is modeled that way. It remains, however, an expensive operation to refresh the ODS. As we have seen in section 3, users often have the requirement to run aggregate queries on line-item level data multiple times a day, which opposes the concept of, for example, refreshing an ODS once a day.

5.2   Latency-Reduced Reporting Architectures

One possible to optimize timeliness of operational data being available in an OLAP environment would be to shorten the intervals between ETL runs to a minimum. The main disadvantage of such *Microbatch* approaches [3] is the resource consumption of the frequent ETL runs: The ETL process should only run in a defined batch window, because the query performance of the warehouse is dramatically affected during ETL processing time.

To enable less resource-intensive ETL processing, architectures have been proposed that move the data transformation outside of the ETL process. Instead, the transformations are done in the warehouse after extraction and loading. Such processing is called ELT, respectively [15]. Also, push architectures for ETL have been proposed in order to replace bulk processing with the handling of deltas on a business or database transaction level, cf. [12, p. 427]. Kimball further suggests to separate historical data from recent data in a warehouse. The recent data is constantly copied into the so-called *real-time partition* using the push approach described above. In doing so, the data warehouse can still be optimized for queries on historical data, while recent events in an enterprise are also recorded in the warehouse. Brobst suggests to extend typical message broker infrastructures in a way that they leverage the ETL push architecture described above [5]. This is done by hooking a data warehouse adapter into the message bus that subscribes to messages which are relevant for the data in the warehouse. Necessary data transformations are done in the warehouse, resembling the concept of ELT, also described above. While the presented approaches come with less data-capture latency than traditional, batch-oriented ETL architectures, changes in the OLTP systems must still be propagated to the data warehouse, where they are consolidated.

The notion of virtual ODS, as opposed to the traditional, physical ODS discussed above, describes a pull-oriented OLAP architecture which gathers the requested information at query run-time. The ODS is virtual in the sense that it translates data warehousing queries into downstream queries to OLTP or third-party systems without persisting any data. Inmon argues that virtual ODS architectures are of limited use when the data in the source systems is not integrated [9]. This is due to the fact that virtual ODS systems do not provide ETL transformations at run-time, which would be necessary to provide for data integration. The reason is that ETL transformations are costly and there is, thus, a trade-off between the extent of functionality offered in a virtual ODS and the response times experienced by end-users. Virtual ODS is the concept which is closest to the virtual cube approach presented in section 4. The difference is, however, that virtual ODSs negatively impact the performance of the OLTP system they re-direct the OLAP queries to. Our virtual cube is designed to perform queries on the column store database part of our hybrid OLTP database, and does therefore not affect the performance of OLTP.

6   Conclusions

In this paper, we have introduced a hybrid architecture for OLTP systems that consists of a row-oriented database for high-volume OLTP inserts, updates and full-row reads for transactions and a column store for reads classified as operational reporting queries.

The column store holds a replica of the data in the row store from which it differs in that

  – only the subset of the OLTP data that is relevant for operational reporting is available and

– the data is transformed into flat (de-normalized) tables when it is transferred into the column store. This transformation step does include neither consolidation for integrating different data sources nor data cleansing as common in a traditional data warehouse setting.

The data in the column store is consistent with the data in the row-oriented database, which we have ensured by extending the transaction handling of the row store to span across the inserts and updates of the column store. The architecture presented in this paper allows for operational reporting, which can be characterized as "lightweight analytics", directly on top of OLTP data. It solves the problem of mixing short-running OLTP queries with long-running (in relation to pure OLTP) operational reporting queries by directing each type of query to the appropriate store (i.e. row or column store) while keeping both stores transactionally consistent. Our prototypical implementation has been commercialized in SAP Business ByDesign since operational reporting seems to be especially useful in mid-market B2B environments. We also presented a virtual cube concept which allows to consume operational reports from outside an OLTP system, namely from a data warehouse. The virtual cube offers a standard cube interface which is mapped to queries against the column store part of our OLTP system. Our prototype of the virtual cube has also been transferred into Business ByDesign.

6.1   Future Work

One possible direction for future research would be to conceptually move the delta index in the column store completely into the row store, instead of buffering the column store delta index in queue tables in the row store. Fractured Mirrors [17] and C-Store [20] provide a row-oriented delta mechanism, but do not propose ACID transactions as we do in the architecture presented here. Another direction for future work addresses the issue that our hybrid approach is currently local to one instance of an OLTP system. While this appears not to be problematic for SMEs, it might be the case that larger corporations want to use operational reporting in an on-premise setting, potentially with the requirement of consolidation across multiple OLTP systems. When adding this requirement, the approach must be extended with general purpose ETL functionality. The ETL process comprises activities such as accessing different source databases, finding and resolving inconsistencies among the source data, transforming between different data formats or languages, and, typically, loading the resulting data into a warehouse. On the case of operational reporting, however, ETL activities should be postponed to query runtime. When doing so, it is probably most challenging to map the transformation steps in ETL processes to operations which can be efficiently computed on-the-fly using column stores. ETL processes model complex data flows between the transformations. Transformation activities can be of *atomic* or *composed* nature. An example from the field of financial accounting would be to show a list containing the opening and closing balance of an account from January to December of a given year: For each month m, an aggregation has to be done on only those line items carrying a date between January 1st and the last day of m. The result is then both the closing balance of

m and the opening balance of $m + 1$. In this example, the sum operator (i.e. the aggregation) would be an atomic transformation. A composed transformation is used to model the process of creating all opening and closing balances. For every complex report, such workflow-like models could be used for describing the transformations. Simitsis, Vassiliadis, and Sellis treat ETL processes as workflows in order to find optimizations [19]. Their research is, however, aimed at optimizing traditional, batch job-like ETL processes. To our knowledge, workflow models for on-the-fly ETL processes have not been investigated. The corresponding research tasks include the identification of complex reporting scenarios and the complex transformations they require. An adequate abstraction must then be found for these transformations, so that they can be generalized to build ETL workflows with them. Then, efficient implementations on a column store must be found for the identified transformations.

Acknowledgments

This project has been done in cooperation with SAP AG. In particular, we would like to thank Torsten Bachmann, Franz Färber, Martin Härtig, Roland Kurz, Gunther Liebich, Yaseen Rahim, Frank Renkes, Daniel Schneiss, Jun Shi, Vishal Sikka, Cafer Tosun, and Johannes Wöhler. We would also like to thank our students, Tilman Giese, Holger Just, Murat Knecht, Thimo Langbehn, Dustin Lange, Mark Liebetrau, Janek Schumann, and Christian Schwarz, for their work.

References

1. Abadi, D.J.: Query Execution in Column-Oriented Database Systems. PhD thesis, Massachusetts Institute of Technology, Cambridge, MA, USA (Feburary 2008)
2. Abadi, D.J., Madden, S.R., Ferreira, M.: Integrating Compression and Execution in Column-Oriented Database Systems. In: SIGMOD 2006: Proceedings of the 2006 ACM SIGMOD international conference on Management of data, pp. 671–682. ACM Press, New York (2006)
3. Adzic, J., Fiore, V., Spelta, S.: Data Warehouse Population Platform. In: Jonker, W. (ed.) VLDB-WS 2001 and DBTel 2001. LNCS, vol. 2209, p. 9. Springer, Heidelberg (2001)
4. Boncz, P.: Monet: A Next-Generation DBMS Kernel for Query-Intensive Applications. PhD thesis, Universiteit van Amsterdam, Amsterdam, Netherlands (May 2002)
5. Brobst, S.: Enterprise Application Integration and Active Data Warehousing. In: Proceedings of Data Warehousing 2002, pp. 15–23. Physica-Verlag, Heidelberg (2002)
6. Chang, F., Dean, J., Ghemawat, S., Hsieh, W.C., Wallach, D.A., Burrows, M., Chandra, T., Fikes, A., Gruber, R.E.: Bigtable: A Distributed Storage System for Structured Data. In: USENIX 2006: Proceedings of the 7th conference on USENIX Symposium on Operating Systems Design and Implementation, Berkeley, CA, USA, p. 15. USENIX Association (2006)
7. Codd, E.F.: A Relational Model of Data for Large Shared Data Banks. Communications of the ACM 13, 377–387 (1970)

8. Copeland, G.P., Khoshafian, S.: A Decomposition Storage Model. In: Navathe, S.B. (ed.) Proceedings of the 1985 ACM SIGMOD International Conference on Management of Data, Austin, Texas, May 28-31, pp. 268–279. ACM Press, New York (1985)
9. Inmon, W.H.: Information Management: World-Class Business Intelligence. DM Review Magazine (March 2000)
10. Inmon, W.H.: Operational and Informational Reporting: Information Management: Charting the Course. DM Review Magazine (July 2000)
11. Inmon, W.H.: Building the Data Warehouse, 3rd edn. John Wiley & Sons, Inc., New York (2002)
12. Kimball, R., Caserta, J.: The Data Warehouse ETL Toolkit: Practical Techniques for Extracting, Cleaning. John Wiley & Sons, Inc., New York (2004)
13. Legler, T., Lehner, W., Ross, A.: Data Mining with the SAP NetWeaver BI Accelerator. In: VLDB 2006: Proceedings of the 32nd International Conference on Very Large Data Bases, pp. 1059–1068. VLDB Endowment (2006)
14. Mahapatra, N.R., Venkatrao, B.: The Processor-Memory Bottleneck: Problems and Solutions. Crossroads 5(3), 2 (1999)
15. Moss, L., Adelman, A.: Data Warehousing Methodology. Journal of Data Warehousing 5, 23–31 (2000)
16. Noaman, A.Y., Barker, K.: A Horizontal Fragmentation Algorithm for the Fact Relation in a Distributed Data Warehouse. In: CIKM 1999: Proceedings of the Eighth International Conference on Information and Knowledge Management, pp. 154–161. ACM Press, New York (1999)
17. Ramamurthy, R., DeWitt, D.J., Su, Q.: A case for fractured mirrors. VLDB J. 12(2), 89–101 (2003)
18. Shukla, A., Deshpande, P., Naughton, J.F.: Materialized view selection for multi-cube data models. In: Zaniolo, C., Grust, T., Scholl, M.H., Lockemann, P.C. (eds.) EDBT 2000. LNCS, vol. 1777, pp. 269–284. Springer, Heidelberg (2000)
19. Simitsis, A., Vassiliadis, P., Sellis, T.: State-Space Optimization of ETL Workflows. IEEE Transactions on Knowledge and Data Engineering 17(10), 1404–1419 (2005)
20. Stonebraker, M., Abadi, D.J., Batkin, A., Chen, X., Cherniack, M., Ferreira, M., Lau, E., Lin, A., Madden, S.R., O'Neil, E., O'Neil, P., Rasin, A., Tran, N., Zdonik, S.: C-Store: A Column-oriented DBMS. In: VLDB 2005: Proceedings of the 31st International Conference on Very Large Data Bases, pp. 553–564. VLDB Endowment (2005)

The Reality of Real-Time Business Intelligence

Divyakant Agrawal

University of California at Santa Barbara
Santa Barbara, CA 93106, USA
agrawal@cs.ucsb.edu
http://www.cs.ucsb.edu/~agrawal

Abstract. Real-time Business Intelligence has emerged as a new technology solution to provide timely data-driven analysis of enterprise wide data and information. Such type of data analysis is needed for both tactical as well as strategic decision making tasks within an enterprise. Unfortunately, there is no clarity about the critical technology components that distinguish a real-time business intelligence system from traditional data warehousing and business intelligence solutions. In this paper, we take an evolutionary approach to obtain a better understanding of the role of real-time business intelligence in the context of enterprise-wide information infrastructures. We then propose a reference architecture for building a real-time business intelligence system. By using this reference architecture we identify the key research and development challenges in the areas of data-stream analysis, complex event processing, and real-time data integration that must be overcome for making real-time business intelligence a reality.

Keywords: databases, data-warehousing, data-cube, data-streams

1 Introduction

In this paper, we take an evolutionary approach to obtain a better understanding of the role of real-time business intelligence in the context of enterprise-wide information infrastructures. The term *business intelligence* dates back to 1958 – introduced by Hans Peter Luhn [5] – an early pioneer in Information Sciences. In spite of this early vision, business intelligence did not evolve into a mainstream technology component until only very recently. The intervening years instead were focused primarily in the development of two key technologies: database management systems (for the most part) and data warehousing. A wide-scale adaptation of DBMS technology led to the proliferation of multiple operational data sources within an enterprise for online transaction processing The need for data warehousing arose in order to consolidate transactional data from multiple operational data sources within an enterprise. The primary role of data warehousing was envisioned as a way to simplify, streamline, standardize, and consolidate reporting systems across the enterprise. In order to cause minimal disruptions to operational data sources, one of the key design decisions was to integrate transactional data in a batched manner.

M. Castellanos, U. Dayal, and T. Sellis (Eds.): BIRTE 2008, LNBIP 27, pp. 75–88, 2009.

The early success of extracting valuable information from historical data stored in large enterprise-level data warehouses accentuated the demand for unlocking *business intelligence*. An enabling technology referred to as the *data cube* [2] further created a wide-scale acceptance of this type of data analysis. Unfortunately, due to the batched approach of data integration, typically there is a significant lag in the currency of information at the data warehouse. In the Internet and world-wide-web context, the role of data warehousing (and business intelligence) has changed considerably from a 2nd class technology component to a critical component in the enterprise information infrastructure. Most commercial enterprises have moved from brick-and-mortar and 8AM-5PM model of operations to the Web-enabled and 24×7 model of operations. The need to track customers on per sales transaction has changed to tracking customers and services on per click interactions. As a consequence the need for performing continuous optimizations based on real-time performance analysis has become a necessity. These requirements warrant us to revisit the traditional DW/BI architectures that are batch-update oriented and are no longer a tenable solution.

In this paper, we propose architectural modifications to enable real-time BI capabilities in DW/BI software stack. We then argue for an event-driven approach to disseminate information from ODSs to the DW storage layer. In doing so, we also introduce a new component referred to as the *stream management layer* which enables a range of stream operators to facilitate real-time analysis of event data. A business-rule engine augmentation at the stream management layer partitions event detection for real-time operational intelligence or for deeper analysis. We conclude this paper with research challenges that need to be addressed before real-time BI can become a reality.

2   The Origins of Business Intelligence

As alluded to in Section 1, the vision of business intelligence was presented in 1958 paper by Hans Peter Luhn [5]. Luhn envisioned a business intelligence system as an automatic system to disseminate information to the various sections of any industrial, scientific, or government organization. Unlike in the current context of business intelligence such as data warehousing and database systems, which are primarily concerned with enterprise-wide transactional data (e.g., click-stream data or sales transaction data), he was primarily focusing on information contained in business-related documents. Thus in his work, Luhn identified three critical technology components for enabling business intelligence: *auto-abstracting* of documents, *auto-encoding* of documents, and *auto-creation* and *updating* of user profiles (action points). The role of his business intelligence system was to *actively* route incoming documents to appropriate users by matching document abstracts with stored user profiles. Figure 1 illustrates the overall architecture of the proposed system.

We make several observations in regard to this vision which is almost 50 years old but still remains valid in the current context. Luhn's original definition of *business* is rather broad. He states "business is a collection of activities carried

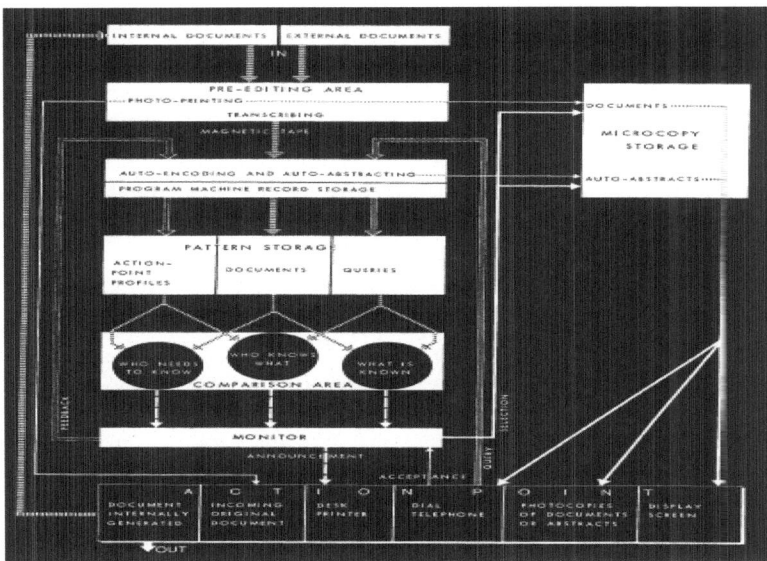

Fig. 1. Luhn's Business Intelligence System for Active Dissemination of Business Documents

on for whatever purpose, be it science, technology, commerce, industry, law, government, defense, etc." In the same vein, he defines *intelligence* in a more general sense "as the ability to apprehend the interrelationships of presented facts in such a way as to guide actions towards desired goal." Finally, we note that the term *business intelligence* predates the notions of *databases* and *data management* which were introduced in the early sixties.

Given that Luhn's notion precedes databases and data management, it is not surprising that his automatic business intelligence system or apparatus used a Microscopy storage system consisting of documents and their abstracts. In some ways, Luhn architected a rudimentary information retrieval system and put forth the idea of *active* information delivery to appropriate action points using his notion of business intelligence (i.e., statistical matching of abstracts to profiles). To a certain degree, it can be argued that the original vision remains relatively unchanged in the current context. However, the underlying information component instead of being a collection of documents is instead enterprise-wide data encompassing diverse types of information. Given the finer granularity of the underlying data and complex interrelationship among different data-elements, our greatest challenge is how to extract useful and meaningful "intelligence" from such data. In the following we now detail the advances and efforts that have been made to overcome this challenge.

3   The Early Years (1970s – 1980s)

In spite the overarching vision of business intelligence which was set forth as far back as in 1958, very little progress was made in the realization of this vision in in the subsequent years – in fact for almost four decades. Instead much of the research and development activities were primarily focused on another nascent technology that was mainly concerned with simplifying the large amount of data processing tasks within an enterprise. Early incarnation of data processing meant running millions of punched cards through banks of sorting, collating, and tabulating machines, with results being printed on paper or punched onto still more cards. Data management meant physically organizing, filing, and storing all this paper and punched cards.

This rudimentary form of data underwent a transformative change with the advent of I/O devices such as magnetic tapes and disks. The basic idea that emerged was to develop data management systems that will enable enterprises to manage their data by storing and cataloging it in a structured manner on such storage devices for subsequent retrieval, update, and processing. In 1961, Charles Bachman at General Electric Company developed the first successful data management system which was termed as *Integrated Data Store* and incorporated advanced features such as data schemas and data logging. IBM subsequently commercialized this technology and marketed its first data management system called IMS in 1968 – which was a hierarchical database that ran on IBM mainframe machines.

Around the same time, IBM researchers in Silicon Valley were actively involved in pushing the frontiers of database technology. Notable among them was IBM researcher Edgar F. Codd who proposed a revolutionary approach for organizing complex enterprise data in terms of flat structures referred to as Tables (or more formally relations). He also, developed a formal declarative language to manipulate data stored in relations and referred to it as relational algebra (also formally called relational calculus). IBM created a large team of researchers during the 1970s to work on a project code-named, System/R, in an effort to commercialize relational databases. Unfortunately, due to sheer inertia of IBM's commitment to IMS and mainframes, IBM did not release a relational database product well into 1980s. Curiously enough, RDBMS became a product in 1973 as a result of two academic researchers Michael Stonebraker and Eugene Wong from UC Berkeley, who founded a company INGRES Corp. which marketed one of the first data management product based on RDBMS technology.

Although IBM can be faulted for not commercializing RDBMS technology in a timely manner, its contributions towards advancing the state-of-the-art of data management technology cannot be ignored. While Codd at IBM was developing the right models and abstractions for representing complex data and information within an enterprise, another group of IBM researchers under the leadership of Jim Gray was working on the orthogonal problem of changing the nature of database from an offline file-processing system to an online transaction processing system. Until then databases were deployed in such a way that daily transactions were processed in a batched environment (generally during the off hours e.g., during the night). This caused a significant latency on the currency of information available from the database especially when it pertained to the current status of important pieces of information such as total amount of cash available in the bank, the current inventory of products on the shelves etc. Jim Gray and his collaborators invented the notion of an *online transaction* as well as developed associated algorithms and system to enable online execution of transactions against the live database to ensure that the database view remains up-to-date at all times. In our view these developments were very critical and perhaps the "seeds" of real-time business intelligence as we see it today. Were it not for online transaction processing we would have remained in the dark ages of information era – in which the current state of the enterprise data would have been clouded by the outdated information in the database.

Latter half of the eighties saw a rapid adaptation of both RDBMS technology and online transaction processing widely in most commercial enterprises especially with the increasing reliance on data and information for both strategic and tactical decision-making.

4   Data Warehousing (1990s –)

Rapid proliferation of online transaction processing systems based on the relational model of data resulted not only such systems being used widely but also resulted in multiple such systems being used within an enterprise. In particular, each department or functional area within a large enterprise such as the Financial Department, the Inventory Control Department, the Customer Relations Department, and so on, for variety of reasons had their own operational data store for their day-to-day operations. At the enterprise level, the reliance on data-driven decision making warranted that online reports from such disparate systems be delivered to the end-users which were mostly business analysts distilling information from these reports to make high-level decision making tasks. Demand for online reporting and online data analysis was further fueled due to some of the enabling technologies at this time: personal computing and computerized spread-sheets that made large-scale number crunching significantly easier. There were two powerful forces that led to an overall realization that the then prevalent information infrastructure cannot sustain the status-quo. One of the problems is that due to the complete autonomy of operational data sources within an enterprise, obtaining a unified analytical view of data and reports derived from multiple system was extremely labor-intensive task. Multiple teams of analysts needed to manually construct a unified view over reports derived from multiple data sources – a task which was highly prone to errors and mistakes. The second problem with this infrastructure was that the ever-increasing demands from business analysts for more reports and data from operational systems started to interfere with the normal day-to-day operations of most departments. This was particularly so – since the workload characteristics of analytical queries typically involved large table-scans, data aggregation and summarization which often resulted in consuming precious I/O and CPU time from normal operations.

In 1988, Devlin and Murphy [1] reported a design of an integrated business information system that was developed to create a unified view of data and information contained in multiple operational systems that were in use by different functional areas within IBM. In this paper, the authors report that the overall architecture of such a system was motivated due to the *fractured* view of an enterprise since data accessors needed to request reports from multiple operational systems. It is in this article, the authors coined the term *business data warehouse* and propose an architecture shown in Figure 2. Furthermore, the authors report that similar efforts were underway at that time in many other large enterprises where there was an emergent need to create an integrated view of business data and information stored in multiple operational data sources.

From the technology market perspective, there were several companies such as Teradata (founded in 1979) there were primarily focused on marketing RDBMS technology primarily for decision support systems. Such technology companies reoriented themselves as data warehousing providers and the next few years saw an increasing interest in building enterprise scale data warehouses. The early years of adaptation of data warehousing was mired with severe problems. Early approaches of data warehouse design relied on a rather ad-hoc design process. This design process entailed bringing schema-level information from multiple operational data sources and then a painstaking efforts were needed to resolve all these schemas with each other to create one giant schema at the global level.

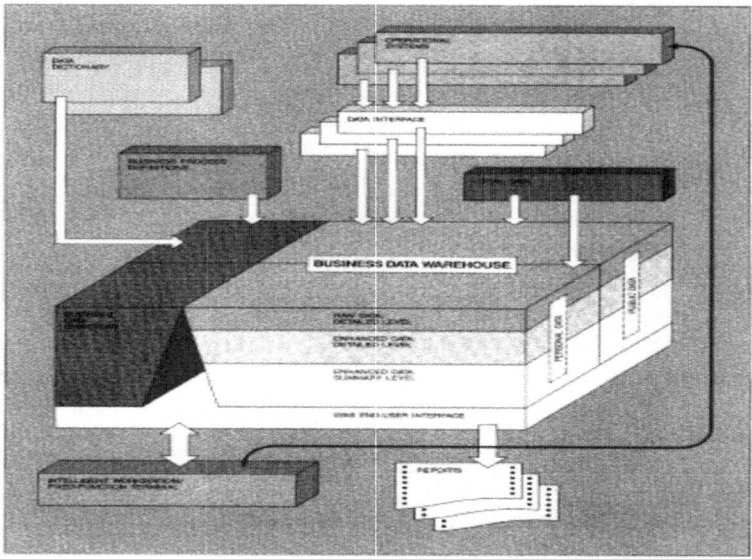

Fig. 2. The Business Data Warehouse Architecture used for IBM's Business Information System (1988)

This design approach was not only very costly but also very time consuming. The whole process was further bogged down with the complexity of constant schema changes that occurred at the operational data sources. Thus the early efforts for building large-scale data warehouses met with very little success. From the management perspective, data warehousing was viewed as unproven, immature, and expensive technology proposition. The distinction between DBMS and Data Warehousing was not clear and seen as duplication of data at multiple levels. Finally, an important non-technical factor common to many failed data warehousing projects was that there was no clear owner or stake-holder of the enterprise data warehouse resulting in the project itself being relegated to a second-class entity.

It was only in the mid-nineties, when two database technology practitioners articulated a methodical approach for designing data warehouses [3,4]. Inmon [3] articulated a systematic approach of applying *database normalization* to all collected schemas using a top-down approach. Kimball [4], on the other hand, championed the case for *dimensional model* and *star schemas* arguing that both from analysis perspective as well as query processing performance perspective, dimensional design is the best approach. This debate resulted in data warehousing efforts to focus on design methodology instead of being focused on and enamored with technology. In particular, IT professionals realized that data warehousing cannot be bought as a shrink-wrapped product. Instead, methodical approaches are needed to gather user requirements from different constituencies to build the dimensional model as well as the individual dimensions at the right granularity. Over the last few years, the Kimball approach of data warehouse design has become the de-facto standard.

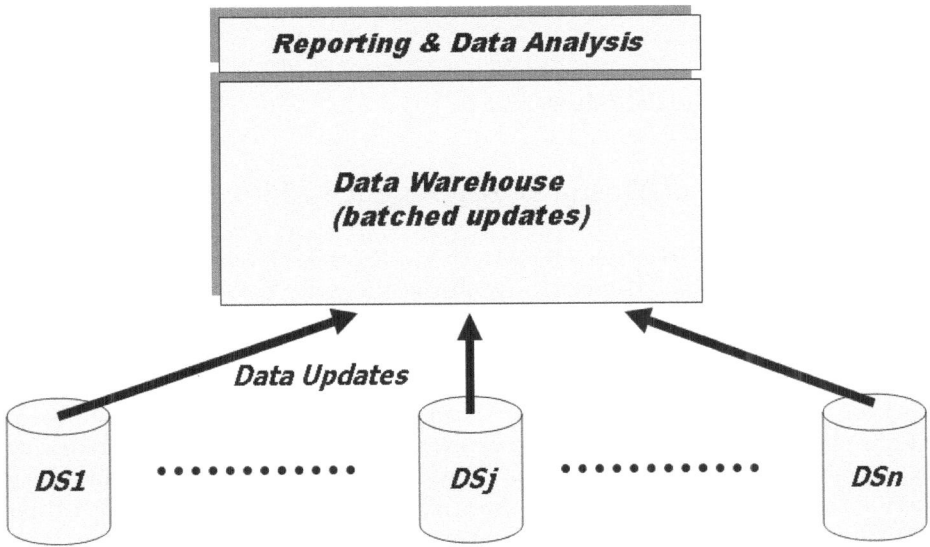

Fig. 3. An Architecture for a Data Warehouse

The conceptual architecture of a data warehouse is depicted in Figure 3. As shown in the Figure, the conceptual architecture is fairly straightforward in that it stipulates integration of data derived from multiple data sources needs to be installed as an integrated data warehouse view. Some of the difficulties that arise in the realization of this architecture is the component software to extract data from the data sources, transform this data so that it is aligned with the data warehouse schema, and loading of the transformed data. This procedure is referred to as ETL process (Extraction, Transformation and Loading). As far as the implementation of the data warehouse itself is concerned, the underlying technology supporting the data at the data warehouse remains basically the relational data model. In spite of this commonality, the technology vendors continue to distinguish themselves as being Database vendors (e.g., Oracle) or Data Warehouse vendors (e.g., Teradata). The key distinction that arises is that database technology is highly optimized for transaction processing whereas data warehouse technology is primarily designed to for running complex queries involving whole-table scans, joins, and aggregation. In general, parallel hardware is extensively used in the context of data warehouses to achieve the desirable performance.

In order to diffuse the tension between DW architects and DBAs running the operational data sources, Kimball articulated batched updates of the data from the ODSs to the DW. This resulted in data warehouses that do not provide up-to-date "state" of the enterprise. Recently, with increasing reliance on the data from data warehouse for high-level decision-making, it is being felt that data warehouses should be updated in real-time and therefore the term *real-time data warehousing*. However, real-time updating of data-warehouses gives rise to host of challenges – one of them being maintaining the consistency of the warehouse view. In conclusion, the key transformative factor for increasing acceptance of data warehousing is due to the notion of *dimensional model based design methodology* – it considerably simplifies the data warehouse design process resulting in both cost and time savings.

5   Emergence of Business Intelligence (2000–)

By the late nineties and early 2000, Data Warehousing found ready acceptance in the business arena. Although early justifications for data warehousing were primarily driven by the needs to provide integrated reporting functionality, the value of data warehousing became clear for carrying out large analysis tasks to assist data-driven decision making both for tactical and strategic management decisions. As the role of analysis expanded rapidly within an enterprise, teams of business analysts within an enterprise were involved in extracting "interesting patterns" from enterprise wide data. This notion of extracting and unlocking useful information from raw data was termed as *business intelligence*. A companion technology that emerged around this time was referred to as *data mining* which leveraged statistical techniques and machine learning algorithm to enable sophisticated analysis over vast amounts of raw data. For example, data mining techniques were rapidly adapted for marketing by analyzing historical behavior of customers' buying patterns. Furthermore, the notion of *business intelligence* became more relevant especially in the context of Web-based commercial enterprises which were able to collect not only the sales-based information of its customers but were also able to collect browsing (or click-level) behavior of its customers. Business intelligence was now seen as a crucial technology component that could be used to make sure that a majority of customer visits to an E-commerce site will result in a *conversion* to a sales activity. Although we give motivation for business intelligence in the context of a store-front – its role can easily be generalized to other activities of an enterprise: inventory control, customer relation management, order processing, and so on.

Although the case for deriving business intelligence from enterprise data was well articulated by the technology marketeers, it was primarily used as a catalyst to sell the concept of data warehousing to the top-level management. In particular, no attempt was made to clarify if there is a key technology component that enables unlocking of "business intelligence" from enterprise wide data stored at the data warehouse. Typical approach was to develop custom-made data analysis applications for aggregating data across many dimensions looking for anomalies or unusual patterns. The main shortcoming of these approaches was they relied on SQL aggregation operators in that the SQL aggregate functions and the GROUP BY operator produced zero-dimensional or one-dimensional aggregates. Business intelligence applications instead need the N-dimensional generalization of these operators. Gray et al. [2] developed the notion of such an operator, called the data cube or simply cube. The cube operator generalized the histogram,

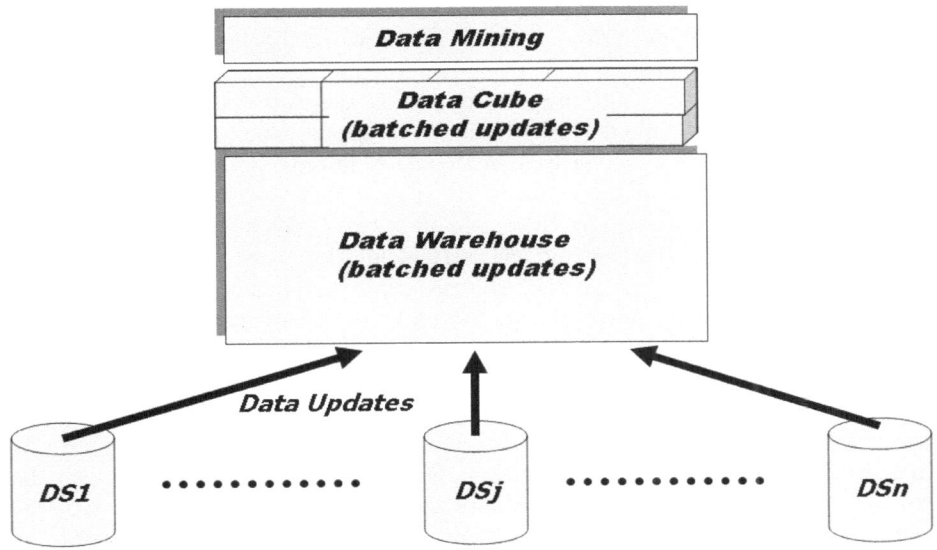

Fig. 4. An Architecture for a Business Intelligence System

cross-tabulation, roll-up, drill-down, and sub-total constructs that were needed for business intelligence activities. The novelty of this approach is that cubes maintained the relational framework of data. Consequently, the cube operator can easily be embedded in more complex non-procedural data analysis programs. The cube operator treated each of the N aggregation attributes as a dimension of N-space. The aggregate of a particular set of attribute values is a point in this space. The set of points forms an N-dimensional cube. Super-aggregates are computed by aggregating the N-cube to lower dimensional spaces. This proposal should be considered as a fundamental breakthrough for enabling data-analysis which now legitimately can be termed as "business intelligence". The resulting architecture of business intelligence system is depicted in Figure 4. The two key technology components that form the core of business intelligence systems are the data cube structure and data mining framework that embeds the cube operator for extracting relevant data in the requisite format. This generalization led to the commercialization of the data cube technology which became an integral part of all major data warehousing vendors: Hyperion (now part of Oracle Corp.), COGNOS (now part of IBM Corp.), and Microsoft Analysis Services. Note that the underlying layers still maintained the offline or batched nature of updating the data in the warehouse thus not providing "business intelligence" that is current with respect to the state at individual operational data sources.

6   Real-Time Business Intelligence

Although the technology landscape for business intelligence is still evolving, there is an emerging debate among database researchers and practitioners about the need for *real-time business intelligence*. The argument put forth by the proponents is that it is not enough to deliver "business intelligence" to the decision-makers but it should also be done in a "timely" manner. This notion of timeliness has resulted in the wide usage of the term *real-time* business intelligence. In a recent article, Schneider [8] provides numerous examples in the context of click-stream analysis to clearly identify the requirements of "degree of timeliness" in a variety of analysis tasks. He argues that not every analysis task warrants real-time analysis and the trade-off between the overhead of providing real-time business intelligence and the intrinsic need for such an analysis should be carefully considered. Otherwise, the resulting system may have prohibitive cost associated with it. Nevertheless, the real-time BI proponents present several compelling examples derived from actual case-studies where it can be shown that real-time BI can provide significant benefits to an enterprise. We envision that as the role of enterprises become increasingly real-time (e.g., E-commerce enterprises that operate 24×7), real-time BI will indeed play an increasingly important role in the routine operations of such companies. In particular, the role of real-time BI will serve as an immediate (and automated) feedback to the operational data sources for making online tactical decisions. For example, the window of opportunity for up-sell or cross-sell a product is while a customer is still around on an E-commerce site not after he/she has left. Thus the need for real-time BI is very well justified. Assuming this our goal is to clearly define the system architecture for a real-time BI system and identify the key technologies that must be developed to make the vision of real-time BI a reality much in the same way as was done in the context of data warehousing (with dimensional modeling) and business intelligence (with the data cube operator).

Our proposal for a real-time business intelligence architecture is depicted in Figure 5. As shown in the figure, we leverage from the existing architecture for a business intelligence system with two major differences. First is the data delivery from the operational data sources to the data warehouse must be in real-time in the form of what we refer to as data-streams of events. Thus we can no longer rely on batched or offline updating of the data warehouse. Second we introduce a middleware technology component that is referred to as *stream analysis engine.* Thus before the incoming data is integrated with the historical information for integrated reporting and deep analysis it should be subjected to an immediate analysis for possible outliers and interesting patterns. The goal of this analysis is to extract crucial information in real-time and is delivered to appropriate action points which could be either *tactical* (e.g., incorporating as an online feedback to the operational data sources) or *strategic* (e.g., immediate high-level management decision). In the following, we identify some of the key research challenges in realizing such an architecture.

It has been widely recognized that ETL technology needed to load all the data into a data warehouse is a formidable task. During the past few years, several ETL tools have appeared in the marketplace that make this task relatively easier. Nevertheless, in general most of the current ETL technology relies on

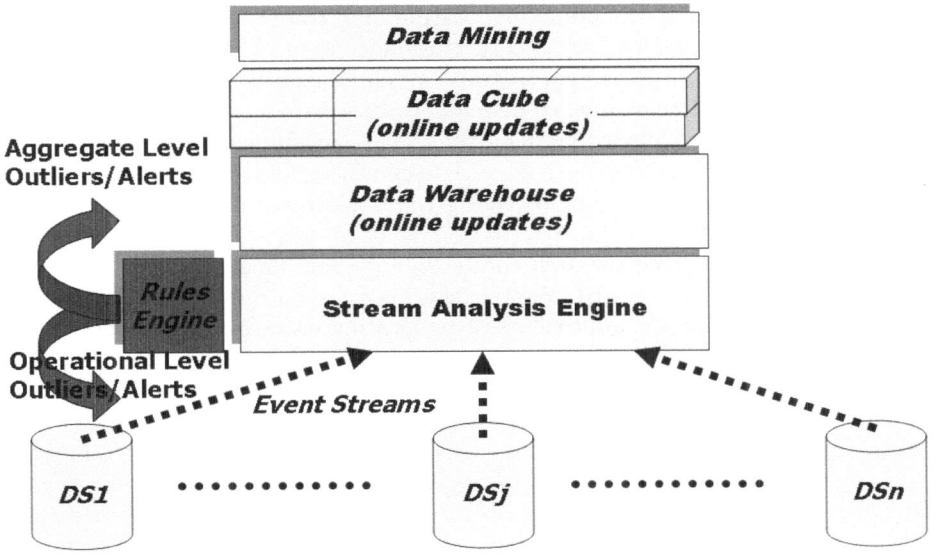

Fig. 5. An Architecture for a Real-time Business Intelligence System

batched updates to the data warehouse. In order to make real-time BI a reality, the ETL process needs to move away from periodic refreshes to continuous updates. Online updating of data warehouses gives rise to numerous challenges. One of the problems that must be dealt with is the overlap of long-term analysis queries in the presence of concurrent updates. The other problem is when data warehouse views are composed on the basis of data derived from multiple data sources, indiscriminate updating can give rise to inconsistency in the views. View synchronization is a complex research topic and efficient solutions are needed to address this problem. Anther complexity that arises is the problem of schema updates and evolution both at the data warehouse as well as at the data sources. Although some progress has been made in this arena, numerous challenges remain. In the same vein, real-time updating gives rise to several challenges with regard to the data transformation that needs to occur before data is loaded into the warehouse. Recent research [6,9] is emerging to formulate the problem of real-time ETL and develop novel solutions.

Our vision of stream analysis engine is to enable event monitoring over incoming data to enable outlier detection and pattern analysis for operational intelligence. We also envision that the stream analysis engine will leverage from models constructed from historical data. For example, in the context of Web search application, this analysis engine will use the historical data from the data warehouse to construct sophisticated user models and query models. Based on this models, incoming event stream can be used to determine if the operational systems are deviating from the norm. Such information can be extremely useful for online tuning of the live operational site. The key research challenge in data stream analysis is that the analysis needs to be done with a single pass over the incoming data. Numerous data-stream operators have been proposed in the data-stream research literature to facilitate online computation of frequent element computation, fraud-detection, performance monitoring, histograms, and quantile summaries. However, considerable research and development efforts are needed to implement these operators in a holistic stream management system. A complementary technology component that will be useful with a stream analysis sub-system is a rule processing engine that can incorporate high-level business rules to enable complex event processing to enable real-time analysis of event stream data.

There are several other research and technology components that are relevant in the context of real-time BI. First is the need for automated techniques for data integration. Current approaches of integrating data from operational data sources into the data warehouse is too tedious and time consuming. There is some new research in the context of Web-based data that enables automated schema integration [7]. We envision that some of these techniques can be useful in the context of real-time BI systems although the proposed system does give rise to the problem of uncertainty of data integration. The second research challenge is the need for new languages and systems for analytical processing. Although SQL has served reasonably well for online analytical processing – it does suffer from significant performance problems. New approaches are being used especially in the context of click-stream data analysis, namely, MapReduce from Google, PigLatin from Yahoo!, and DRYAD from MicroSoft. However, it remains to be seen if these paradigms and languages can indeed be useful for real-time BI. Finally, the most formidable challenge in the case of large scale data warehouses and business intelligence systems to be able to scale with increasing volumes of data. Just a few years ago, we used to talk about data warehouses in the scale of terabytes and now a petabyte-scale data warehouse is not too far-fetched. With this scale of data, certain queries especially those that involve spatial and temporal correlations need to process vast amounts of data. Given the fact that complete scan of data is necessary (perhaps multiple times) to execute such queries, the only recourse we have is to use a data partitioning approach coupled with parallel DBMS technology. Companies such as Teradata have historically used a combination of proprietary parallel hardware and parallel software solutions. Recently, several new vendors (e.g., GreenPlum and Aster Data Systems) have emerged who provide parallel DBMS technology based on commodity hardware and software components. However, emergence of utility computing or cloud computing is likely to bring about new research and development challenges as data management activities are outsourced to external parties.

7 Concluding Remarks

Real-time business intelligence has emerged as a new technology solution to serve the data-driven decision making needs of contemporary enterprises. Real-time BI is likely to play a key role in delivering analytics both for tactical as well as for strategic decision making. Unfortunately, adaptation of real-time BI solutions is hampered since there is no clarity about the underlying technology components and custom solutions are not desirable since they are prohibitively expensive. In this paper, we have taken an evolutionary approach to understand the reasons for early failures and subsequent successes in the context of data warehouses and traditional business intelligence systems. We draw from this experience to clearly outline the reference architecture for a real-time business intelligence system. We then identify key research and development challenges that must be overcome to build the necessary technology components that will serve as the key enablers for building real-time BI systems.

References

1. Devlin, B.A., Murphy, P.T.: An Architecture for a Business and Information System. IBM systems Journal (1988)
2. Gray, J., Chaudhuri, S., Bosworth, A., Layman, A., Reichart, D., Venkatrao, M., Pellow, F., Pirahesh, H.: Data Cube: A Relational Aggregation Operator Generalizing Group-by, Cross-Tab, and Sub Totals. Data Min. Knowl. Discov. 1(1), 29–53 (1997)
3. Inmon, W.H.: Building the Data Warehouse. John Wiley & Sons, New York (1996)

4. Kimball, R.: The Data Warehouse Toolkit: Practical Techniques for Building Dimensional Data Warehouses. John Wiley & Sons, New York (1996)
5. Luhn, H.P.: A Business Intelligence System. IBM Journal of Research and Development (1958)
6. Polyzotis, N., Skiadopoulos, S., Vassiliadis, P., Simitsis, A., Frantzell, N.: Supporting Streaming Updates in an Active Data Warehouse. In: Proceedings of the 23rd International Conference on Data Engineering: ICDE (2007)
7. Sarma, A.D., Dong, X., Halevy, A.Y.: Bootstrapping pay-as-you-go data integration systems. In: Proceedings of the ACM SIGMOD International Conference on Management of Data, SIGMOD (2008)
8. Schneider, D.: Practical Considerations for Real-Time Business Intelligence. In: Bussler, C.J., Castellanos, M., Dayal, U., Navathe, S. (eds.) BIRTE 2006. LNCS, vol. 4365, pp. 1–3. Springer, Heidelberg (2007)
9. Vassiliadis, P., Simitsis, A.: Near Real-time ETL: Annals of Information Systems: New Trends in Data Warehousing and Data Analysis, vol. 3. Springer, US (2008)

Beyond Conventional Data Warehousing — Massively Parallel Data Processing with Greenplum Database

(Invited Talk)

Florian M. Waas

Greenplum Inc., San Mateo CA 94403, USA
flw@greenplum.com
http://www.greenplum.com

Abstract. In this presentation we discuss trends and challenges for data warehousing beyond conventional application areas. In particular, we discuss how a massively parallel system like Greenplum Database can be used for MapReduce-like data processing.

Keywords: Data Warehousing, analytics, petabyte-scale, data processing, MapReduce, User-defined Functions.

1 Introduction

Hardly any area of database technology has received as much attention as the data warehousing segment recently. A number of start-up companies have entered the market with database products specifically engineered for large-scale data analysis. While the discipline goes back to the early 1960s it was only in the 1990s that enough compute resources were readily available to apply the principles of data analysis to significant amounts of data. However, in the past 5 years new application scenarios—often in a Web 2.0 context—have pushed the limits of conventional solutions. Not long ago data warehouses of a total size of 1 Terabyte (TB) were considered the high end of data management from a data volume point of view. Today's data collection mechanisms easily exceed these volumes, however. In our work with customers we frequently encounter application scenarios with daily loads in excess of 1 TB and requirements for several 100s of TB for the entire data warehouse. Until a few years ago, only very few database vendors provided solutions for this market segment, at a price-performance-ratio that was unaffordable for young Web 2.0 companies.

When first confronted with the problem of having to manage massive amounts of data, most of these companies invested in building special purpose solutions. Unfortunately, many of these solutions were not amenable to integration with 3rd party Business Intelligence Tools and lacked many of the traditional database features necessary for advanced analytical processes. At the same time many of these companies demonstrated in impressive ways the value of massive data

M. Castellanos, U. Dayal, and T. Sellis (Eds.): BIRTE 2008, LNBIP 27, pp. 89–96, 2009.

collection and monetization of social networks. This in turn attracted more and more businesses to explore data analytics which in turn drives an increasing demand for standard solutions in this space.

In this presentation, we first discuss the requirements for a new kind of data warehouse and details how the Greenplum Database data management platform addresses them. Then, we focus on advanced analytics and in particular how a massively parallel data processing system opens up a number of opportunities to do much more than simply query data. We conclude with a few directions for future work.

2   Explosive Data Growth

In recent years data warehousing has seen a number of new application areas and customer scenarios. Generally, we witnessed a trend away from the core scenario of analysis of sales data, toward analysis of data that describes behavioral aspects of a clientele rather than their financial transactions. The most prominent examples for these new types of data sources are click-streams or ad-impression logs, often referred to as *behavioral data*.

Unlike traditional sales data, behavioral data does not capture a deliberate decision to purchase goods or make otherwise financial transactions; this has two immediate implications for the analysis of this type of data:

1. individual records are of "lesser" value—the decision to visit a website is typically made more lightly than purchase decisions—which means the analysis needs to take into account large amounts of data to obtain statistically significant insights, and,
2. substantially larger amounts of data than can be processed with conventional means are being produced.

Both these demands have been driving the scalability requirements for modern data warehouses. Many of our customers are constantly refining their tracking processes which results in even larger amounts of data.

Another dimension of data growth is the increasing market penetration of high-tech devices in emerging markets. Sustained growth rates of over 100% for entire industries such as telecommunications in various markets in Asia have been fairly common in the past 5 years. The current developments point to further increases at similar rates in the next years.

Most importantly, the currently observed growth rates continue to out-pace Moore's Law, that is, the advancements in technology development will not catch up with the market trends in the foreseeable future. In other words, we cannot rely on hardware improvements to close the gap between today's conventional database technology and the requirements for large-scale data warehousing solutions any time soon.

Taking the growth expectations for their businesses into account scalability requirements call for storage and analysis at a petabyte scale in the near future.

These requirements are clearly outside the range of what can be achieved with conventional database technology today.

3   A New Kind of Database System

Over the past 5 years, a number of database startups entered the database market with the promise to address the needs for large-scale data warehousing indicating a change in the competitive landscape of database technology.

There have been a few notable attempts by startups prior to this recent surge; mostly catering to special purpose data processing needs. What is different with the current trend is that the new players are emphasizing many of the core competencies of traditional database technology yet pushing the limits far beyond what was until recently state of the art.

Companies like Greenplum address the aforementioned drawbacks of conventional DBMS, most importantly: scalability, performance and fault-tolerance. Unlike traditional database vendors, Greenplum focuses on DW/BI applications exclusively: Greenplum Database is a highly scalable, fault-tolerant, high performance database system with special emphasis on query processing for OLAP. Greenplum Database leverages a shared-nothing architecture using commodity hardware. A typical installation consists of 10's of individual nodes, each featuring 4, 8 or more CPU cores, 16 GB of main memory and up to 40 TB direct-attached storage. Configurations range anywhere from 10's to 100's of terabytes of usable storage.

Greenplum Database addresses a fairly varied market. Many of our customers replaced existing database systems as they reached scalability limits. Hence, one important requirement is for Greenplum Database to be a drop-in replacement and as such integrate easily with existing 3rd party tools both for data loading as well as for analytics and reporting.

In Figure 1 the architecture of Greenplum Database and its integration are depicted. As shown in the figure, Greenplum Database is made up of a configurable number of data hosts and a master host which coordinates query processing as well as administrative tasks. All data is mirrored across hosts for fault-tolerance. Similarly a second master host serves as a stand-by (not shown in the drawing). For brevity's sake we skip a detailed in-depth description of the system and refer readers to technical whitepapers [2,3].

One feature that deserves special mention and is relevant for our later discussion is that of parallel loads: in addition to the standard access through an SQL interface which channels data through the master node, Greenplum Database also provides parallel load capabilities which allow every data host to access external data be it in the form of flat files on file servers or through special loader interfaces that directly connect to source applications. This mechanism can be used not only for loading data into regular data tables in the database but also to integrate data and make it accessible to queries without actually loading it into the database. The high data rates make the latter appear seamless and mitigate the otherwise significant threshold for data integration at a multi-terabyte scale.

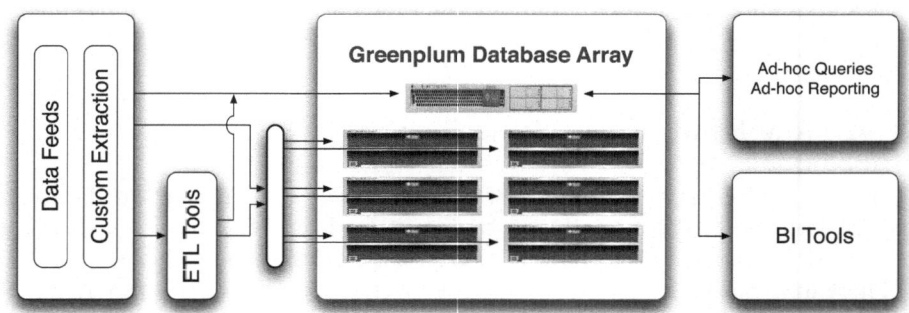

Fig. 1. Architecture and integration of Greenplum Database

4 Alternate Application Areas

Besides providing increased query processing performance, we realized the system could be utilized in other application scenarios too. Given its large number of CPUs and size of main-memory a system can be viewed as a tightly coupled grid computing platform. In particular, in many cases the Greenplum Database system is the single largest system in the data center. So, it was natural to look for alternate processing tasks customers could use the system for. In this presentation we discussed a case study of how using Greenplum Database enabled customers to streamline their data integration processes by moving ETL functionality from a dedicated ETL server farm onto Greenplum Database vastly improving their load times and reducing hardware and maintenance costs for an additional ETL platform.

The idea of using a database system to implement ETL—or ELT—processes has been contemplated previously. However, the high requirements for CPU time and memory limited any attempt to small data sizes and short-running jobs. With the advent of high-powered data warehouse platforms this approach has finally become viable.

Example. The following example illustrates this technique. Using a system of 40 data nodes with a combined 160 CPU cores and a total of 1 TB distributed main memory a daily data loads of 4 TB on average had to be accomplished. Prior to loading, the data was processed by a hand-crafted parallel data cleansing mechanism executed on 6 separate machines. The cleansing of one day's worth of data took typically over 15 hours. Unfortunately due to frequent changes in business requirements data loads have to be repeated to allow cleaning algorithms to be refined etc. From an operational point of view this meant that occasional repeats of daily loads could put daily processing at risk and cause backlogs of daily loads to build up. After a careful analysis of the data cleansing process we re-engineered the mechanisms using only SQL and instead of loading only 'clean' data decided to load raw data directly into staging tables and process that data using SQL before storing it at its final destination. Compared to the original way of processing the data, this approach comes with several benefits:

1. Expressing data transformations as queries allows us to leverage Greenplum Database's query processor infrastructure that parallelizes queries automatically and executes them efficiently leveraging a potentially massive degree of parallelism;
2. Using SQL instead of a scripting language simplified the approach and lead to a cleaner and more robust design;
3. The overall running time for daily loads was reduced to about 2 hours;

Effectively we converted a classical ETL approach into a ELT strategy where load and transformation phase are swapped. As pointed out before, the idea to load raw data and process the transformations inside the database has been considered before, however, only through a massively parallel platform and automatic parallelization of the application does it become viable.

5 Beyond Conventional Data Warehousing

In opening our platform for unconventional use we enable customers to execute data-centric tasks on a parallel high-performance computing platform.

5.1 Declarative Programming Models

SQL provides a powerful, declarative programming model, which enables automatic parallelization to a much larger degree than imperative languages. However, we came to realize that writing complex programs in SQL is widely regarded as a non-trivial task that requires an advanced skill set. In our interaction with customers we found users belong typically to one of two camps: (1) programmers who prefer declarative programming and are comfortable with expressing even complex data processing logic using SQL and (2) programmers who come from a purely imperative background who find SQL difficult to use and non-intuitive. Like most discussions around preferences regarding programming languages, this has been subject of much debate and while certain application scenarios seem to favor one or the other there is no definitive consensus among programmers.

5.2 Other Programming Models

In order to provide the same benefits SQL programmers enjoy with automatic parallelization of their programs etc. to others as well, we investigated alternative programming language paradigms.

In particular, the concept of *User-defined Functions (UDF)* can be used to encapsulate functionality written in a hosted language such as Python, Perl, C, to name just the most prominent ones. This allows users to strike a convenient balance: use SQL to describe data access and simple predicates and combine it with a imperative programming language to express logic otherwise difficult to state as SQL. *User-defined Aggregates (UDA)* are complementary to UDFs and extend the scope of injecting imperative logic in a declarative model to aggregation.

5.3   MapReduce

UDF's and UDA's provide a powerful framework to encapsulate imperative logic and to combine it seamlessly with SQL's declarative paradigm. However, in practice many programmers find it useful to take the abstraction a step further and instead of providing a general framework for using UDX's in a declarative context prefer using scaffolding that conceals the declarative component entirely. This paradigm has become popular through a specific implementation known as *MapReduce* [1] and has been adopted widely by the parallel programming community, see e.g. [5].

We view this programming model as a natural extension to the relational model. Greenplum Database embraces this programming model and provides primitives supporting software development for MapReduce-like operations in its latest version v3.2 [4]. Using its query processing infrastructure, Greenplum Database offers automatic parallelization and mature, highly tuned data access method technology.

5.4   Cost of Data Transfer

As the amount of data increases data becomes more costly to transfer. We found many applications simply extract large amounts of data, transfer it to a client application and perform much of the processing there. While this works fairly well with small amounts of data, this technique becomes prohibitively expensive when large amounts of data are concerned. Having capabilities to express complex logic easily and efficiently for server-side processing is not only an elegant abstraction but becomes increasingly necessary in these scenarios.

6   Future Trends and Challenges

We conclude this presentation by reviewing some of the trends and challenges we anticipate for the near future for a increasingly broader use of database technology in general and data warehousing technology in particular.

6.1   Analytics

The new application areas we mentioned above not only demand new technical solutions to store and access data; they also pose novel challenges to the statistical analysis of data that is of very different nature than data traditionally managed in database systems.

The analysis of behavioral data put a strong emphasis on specific OLAP functionality such as window functions [6] of the SQL language definition. In

addition, customers also demand better integration with special purpose programming languages for complex statistical analysis, e.g. R [3].

We anticipate more and more logic currently implemented in external 3rd
party products to be increasingly moved closer to the data, i.e. integrated with
the data warehouse.

6.2   Hardware Developments

Greenplum Database's success is partly due to the commoditization and availability of servers with significant amounts of direct-attached storage. This is in
stark contrast to most commercial database systems used for transaction processing that deploy large *Storage Area Networks (SAN)* or network-attached storage
systems. In comparison to SAN technology, direct-attached storage lowers the
price-performance ratio significantly making petabyte-scale data warehousing
affordable. The commoditization of direct-attached storage can be expected to
continue at a similar pace. An equally important development is the continuous
advancement of CPU miniaturization resulting in increased numbers of CPU
cores. Greenplum Database's architecture is well-positioned to take advantages
of multi-core CPUs.

Trends toward smaller and less capable, less reliable, lower quality commodity
components as proposed mainly by Web 2.0 companies in the past do not appear
attractive for data warehousing applications due to their high maintenance overhead. Hence, we do not anticipate these to raise to any significant importance
for data warehousing.

6.3   Database Technology and Programming Paradigms

Scalability and high-availability requirements pose interesting challenges as an
increasing number of applications require management of over a petabyte of data.
Similarly, query optimization and execution strategies need to be extended and
adjusted. We anticipate a further departure from the concept of an all-purpose
database as was promoted by most database vendors in the past: many optimizations for executing data warehouse query loads are detrimental for transaction
processing workloads.

7   Summary

In this presentation we detailed some of the challenges for database technology
to achieve petabyte-scale storage and data processing capabilities.

Through extended programming models and mechanisms that enable users to
encapsulate special-purpose programming logic into declarative data processing,
complex data analysis tasks can be carried out leveraging the massively parallel
platform of the data warehouse infrastructure.

References

1. Dean, J., Ghemawat, S.: MapReduce: Simplified Data Processing on Large Clusters. In: OSDI 2004: Sixth Symposium on Operating System Design and Implementation, San Francisco, CA (December 2004)
2. Greenplum Inc., Greenplum Database: Powering the Data-driven Enterprise (June 2008), http://www.greenplum.com/resources
3. Greenplum Inc., A Unified Engine for RDBMS and MapReduce (October 2008), http://www.greenplum.com/resources
4. Greenplum Inc., Release Notes for Greenplum Database 3.2 (September 2008), http://www.greenplum.com/resources
5. Olston, C., Reed, B., Silberstein, A., Srivastava, U.: Automatic Optimization of Parallel Dataflow Programs. In: USENIX 2008, Annual Technical Conference, Boston, MA (June 2008)
6. SQL Standard 2003, SQL/OLAP, Online Analytical Processing: Amendment (2003)

Scalable Data-Intensive Analytics

Meichun Hsu and Qiming Chen

HP Labs
Palo Alto, California, USA
Hewlett Packard Co.
{meichun.hsu,qiming.chen}@hp.com

Abstract. To effectively handle the scale of processing required in information extraction and analytical tasks in an era of information explosion, partitioning the data streams and applying computation to each partition in parallel is the key. Even though the concept of MapReduce has been around for some time and is well known in the functional programming literatures, it is Google which demonstrated that this very high-level abstraction is especially suitable for data-intensive computation and potentially has very high performance implementation as well. If we observe the behavior of a query plan on a modern shared-nothing parallel database system such as Teradata and HP NeoView, one notices that it also offers large-scale parallel processing while maintaining the high level abstraction of a declarative query language. The correspondence between the MapReduce parallel processing paradigm and the paradigm for parallel query processing has been observed. In addition to integrated schema management and declarative query language, the strengths of parallel SQL engines also include workload management and richer expressive power and parallel processing patterns. Compared to the MapReduce parallel processing paradigm, however, the parallel query processing paradigm has focused on native, built-in, algebraic query operators that are supported in the SQL language. Parallel query processing engines lack the ability to efficiently handle dynamically-defined procedures. While the "user-defined function" in SQL can be used to inject dynamically defined procedures, the ability of standard SQL to support flexibility of their invocation, and efficient implementation of these user-defined functions, especially in a highly scaled-out architecture, are not adequate This paper discusses some issues and approaches in integrating large scale information extraction and analytical tasks with parallel data management.

1 Introduction

Scaling data warehouses and analytics to offer *operational business intelligence*, which expands business intelligence into real-time adaptive transactional intelligence, is an important growth opportunity for the BI industry. Next generation business intelligence will have much more sophisticated algorithms, process peta-size of dataset, and require real-time or near real-time responses for a broad base of users. It will also integrate information in near real-time from more heterogeneous sources, which will require processing large quantity of data and applying sophisticated

M. Castellanos, U. Dayal, and T. Sellis (Eds.): BIRTE 2008, LNBIP 27, pp. 97–107, 2009.
© Springer-Verlag Berlin Heidelberg 2009

transformation and summarization logic. Furthermore, the growing demand for fusing BI with data extracted from less-structured content such as documents and the Web is expected to further increase the need for scalable analytics. While massively parallel data management systems (e.g. IBM DB2 parallel edition, Teradata, HP Neoview) have been used to scale the data management capacity, BI analytics and data integration analytics have increasingly become a bottleneck. Figure 1 (left) illustrates the current state of BI architecture and the area of challenge.

1.1 The Challenges

With the current technology, BI system software architecture largely separates the BI analytics layer (BI applications and tools) and the data integration layer (ETL tools and data ingest applications) from the data management layer (DW). The data management layer manages the data in the data warehouse and requests expressed in queries with support of simple aggregates, such as count, sum, min, and max. Many problems are caused by the separation of analytics and data management in this traditional approach:

- The amount of data transferred between the DW and BI is becoming the major bottleneck. This problem has become more pronounced due to the explosion of data volume and the faster progress on CPU than data bandwidth.
- There is growing concern in large enterprises on the security issue raised by moving large quantity of data, some of which sensitive, to various applications outside of the data warehouses.
- When the data set required for analytics is large, once that data set arrives at the BI analytics application layer, the application layer again is burdened with many generic data management issues (data structure, layout, indexing, buffer management etc.) that the data management layer excels at but now need to be duplicated or handled ad hoc at the application layer.
- Larger data sets imply higher compute requirement at the BI analytics layer. However, the opportunity to balance resource utilization between the parallel DWM and parallel BI analytics is lost when these two layers are distinctively separated, as they are now.

Converging data-intensive analytics computation into a parallel data warehouse is the key to address these problems [4-6]. One option for the next generation BI system, illustrated in Figure 1, will have data intensive part of analytics executed in the DW engine.

To effectively handle the scale of processing required in information extraction and analytical tasks, partitioning the data streams and applying computation to each partition in parallel is the key. Google demonstrated that the MapReduce abstraction is very suitable for highly parallel data-intensive computation [2]. The correspondence between the MapReduce parallel processing paradigm and the parallel query processing paradigm has been observed [3]. However, the parallel query processing paradigm has focused on built-in algebraic query operators that are supported in the SQL language. Parallel query processing engines lack the ability to efficiently handle dynamically-defined procedures. While the "user-defined function" in SQL can be used to inject dynamically defined procedures, the ability of standard SQL to support

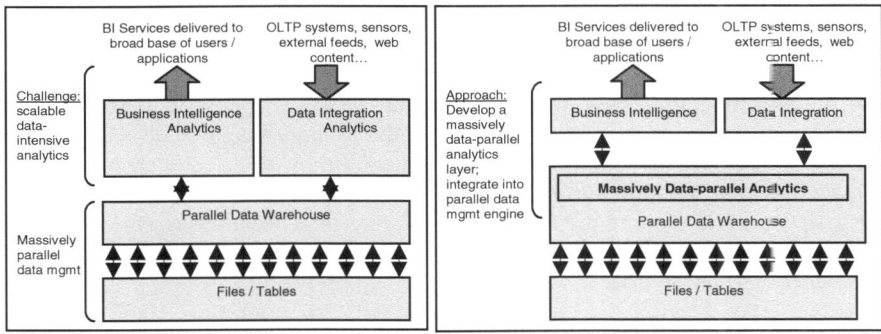

Fig. 1. Analytics and data management separated (left) and converged (right) BI Platform

flexibility of their invocation, and efficient implementation of these user-defined functions, especially in a highly scaled-out architecture, are not adequate.

There has been quite some research over the years that touch on the issue of converging data intensive computation and data management. Integrating data mining with database systems has been explored. Using UDFs is the major mechanism for carrying out data intensive computation inside database systems. Briefly speaking, a UDF provides a mechanism for extending the functionality of a database server by adding a function that can be evaluated in SQL statements. SQL lacks efficient support for content processing primitives, which will require much more flexible and open definitions of the procedures. For example, if one wishes to perform word-count on a text document, the variation in the word-count procedure can be significant depending on the context at hand. For example, the treatment for "stop words" and "noun phrases" may vary a great deal. Wrapping applications as UDFs provides the major way to push computations down to the data management layer; however it also raises some hard issues.

1.2   The Research Directions

We address the above challenges from several fundamental directions.

Extend UDFs Modeling Power to Support Parallel Processing. Existing SQL systems offer scalar, aggregate and table functions, where a scalar or aggregate function cannot return a set; a table functions does return a set but its input is limited to a single-tuple argument. These types of UDFs are not relation-schema aware, unable to model complex applications, and cannot be composed with relational operators in a SQL query. Further, they are typically executed in the tuple-wise pipeline in query processing, which may incur performance penalty for certain applications, and prohibits data-parallel computation inside the function body. Although the notion of relational UDF has been studied by us [1] and others [7], it is not yet realized on any product due to the difficulties in interacting with the query executor.

Construct SQL process with UDFs to Express General Graph based Dataflow. Many enterprise applications are based on the information derived step by step from continuously collected data. The need for automating data derivation has given rise to the integration of dataflow processes and data management in the SQL framework. However, a single SQL statement has limited expressive power at the application level, since the data flow represented in a SQL query is coincident with the control flow, but an application often requires additional data flows between its steps. In other words, the query result at a step cannot be routed to more than one destination, while the requirement of the dataflow of an application is often modeled as a DAG (Directed Acyclic Graph). In order to integrate data intensive computation and data management while keeping the *high-level* SQL interface, we envisage the need for the synthesis of SQL and business processes where the business processes describe the information derivation flow.

In the next Section we describe our work in extending UDF to support MapReduce-like parallel computation inside a parallel database engine. In Section 3 we illustrate our solution on SQL process. In Section 4 we give concluding remarks.

2   Extend UDFs to Support Parallel Processing

In this section we describe the UDFs extensions we are developing to enable rich, MapReduce-style analytics to run on a parallel database system. We will illustrate throughout with K-Means clustering as an example.

2.1   K-Means Clustering Algorithm

The k-means algorithm is used to cluster n objects based on attributes into k partitions, $k \ll n$. It is similar to the expectation-maximization algorithm for mixtures of Gaussians in that they both attempt to find the centers of natural clusters in the data. It assumes that the data attributes form a vector space. The objective is to minimize total intra-cluster variance, or, the squared error function

$$V = \sum_{i=1}^{k} \sum_{p_j \in C_i} (p_j - \mu_i)^2$$

where there are k clusters C_i, $i = 1, 2, ..., k$, and μ_i is the center or mean point of all the points $p_j \in C_i$.

K-Means is an iterative algorithm. Every iteration of K-means starts with a set of cluster centers, and executes two steps: the *assign_centers* step, which takes every point in the set and identifies a center the point is closest to based on some distance function (not necessarily Euclidian), and assigns the point to that cluster; and the *compute_new_centers* step, finds, for every cluster, the geometric mean of all the points assigned to that cluster to be the new centers.

Let us consider the SQL expression of K-Means for two-dimensional geographic points. In one iteration, the first step is to compute, for each point in relation *Points* [point_id, x, y], its distances to all centers in relation *Centers* [cid, x, y]), and assign its membership to the closest one, resulting in an intermediate relation of the form *Nearest_centers* [x, y, cid]. The second step is to re-compute the set of new centers based on the average location of member points. In SQL, these two steps can be expressed as

[Query 1: K-Means with conventional scalar UDF]

```
SELECT Cid, AVG(X), AVG(Y) FROM
    (SELECT P.x AS X, P.y AS Y,  (SELECT cid FROM Centers C WHERE
    dist(P.x, P.y, C.x, C.y) = (SELECT MIN(dist(P2.x, P2.y, C2.x, C2.y))
    FROM Centers C2, Points P2 WHERE P2.x = P.x AND P2.y = P.y) )  AS Cid
    FROM Points P)
GROUP BY Cid;
```

Since this query uses a scalar UDF evaluated on the per-tuple basis, and the UDF is unable to receive the whole Centers relation as an input argument, the Centers relation is not cached but retrieved for each point, and the Centers relation is retrieved again in a nested query for each point p, to compute the MIN distance from p to the centers. Such relation fetch overhead is caused by the lack of relation input argument for UDFs. From the query plan it can be seen that the overhead in fetching the Centers relation using scalar UDF is quite excessive. Such overhead is proportional to the number of points.

In addition to the problem of retrieving input relations repeatedly, some applications cannot be modeled without the presence of whole relations (such as minimal spanning tree computation). Further, feeding a UDF a set of tuples rather than a single one is critical for data-parallel computation using multi-core or GPU. All these have motivated us to support Relation Valued Functions (RVFs).

2.2 K-Means Using Relation Valued Function

The conventional scalar, aggregate and table UDFs are unable to express relational transformations since their input or output are not relations, and hence cannot be composed with other relational operators in a query. The conventional UDFs are typically processed with tuple-wise input which may incur modeling difficulty or performance penalty for some applications [7]. In order to overcome these limitations we introduce a kind of UDFs with input as a list of relations and return value as a relation, called Relation Valued Functions (RVFs). An RVF derives a relation (although it can have database update effects in the function body) just like a standard relational operator, thus can be naturally composed with other relational operators or sub-queries in a SQL query, such as

$$SELECT * FROM RVF_1(RVF_2(Q_1, Q_2), Q_3);$$

For example, a single iteration of K-Means clustering can be expressed by the following query that invokes an RVF, where the RVF receives data from the *Points* relation tuple by tuple, and from the *Centers* relation as a whole.

[Query 2: K-Means with RVF]

```
SELECT Cid, avg(X) AS cx, avg(Y) AS cy
FROM
(SELECT p.x AS X, p.y AS Y,
    nearest_center_rvf (p.x, p.y, "SELECT cid, x, y  FROM Centers") AS Cid FROM
Points p)
    GROUP BY Cid;
```

In this query, as the repeated retrieval of Centers data is avoided, it has much better performance.

2.3 K-Means in MapReduce Style on Parallel DB

A MapReduce scheme consists of two operations: map() and reduce(). The map() reads a set of "records", does any desired filtering and/or transformation, and then outputs a set of records of the form (key, data). Then a "split" function partitions the records into M disjoint buckets by applying a function to the key of each output record. This split function is typically a hash function, though any deterministic function will suffice. In general, there are multiple instances of the map() operation running on different nodes of a compute cluster; each map() instance is given a partition of the input data. The key thing to observe is that all map instances use the same hash function. Hence, all output records with the same hash value will be in corresponding output buckets.

The second phase of a MapReduce job executes M instances of the reduce(). All output records from the map phase with the same hash value will be consumed by the same reduce instance, no matter which map instance produced them.

To implement and execute one iteration of the K-Means clustering algorithm based on MapReduce, it would look like what is shown in Fig. 2. It starts with a set of cluster centers, and executes in two phases:

– finding nearest_centers corresponds to a *map* function;
– computing new centers corresponds to a *reduce* function.

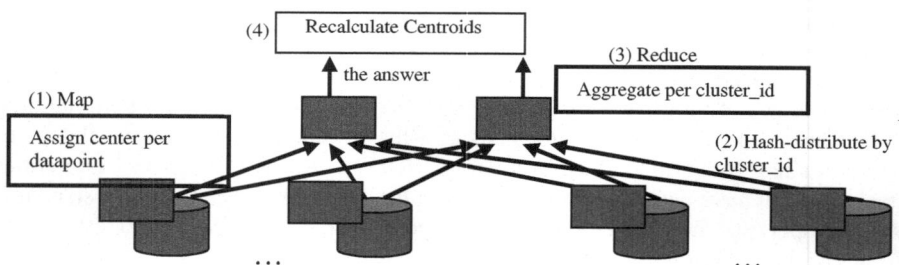

Fig. 2. MapReduce Execution of One Iteration of K-Means

On a parallel database engine, while the scalar UDF-based **Query 1** fails to exhibit a similar execution pattern as MapReduce, our RVF-based **Query 2** behaves with essentially the same parallel execution pattern and data flow. In **Query 2**, *map* is like the clause for generating *Nearest_centers* [x, y, cid], and *reduce* is analogous to the aggregate function (in this case, AVG()) that is computed over all the rows with the same group-by attribute. On a parallel database, the *Points* table is hash partitioned by point_id over multiple server nodes, and the map function for assigning the nearest centers is applied to each data partition in parallel, each yielding a local portion of the result-set, *Nearest_Centers* [x, y,cid], which is automatically re-partitioned by center id (cid) for the parallel *reduce* computation as

SELECT cid, AVG(x), AVG(y) FROM nearest_centers GROUP BY cid

On a parallel database, aggregating a query result group by selected attributes, is implemented exactly in the MapReduce style, except the operators corresponding to map and reduce are not arbitrary computations.

3 SQL Query Process

3.1 Handle Enterprise Dataflow Inside Database Engine

Enterprise applications often implement the information flow and derivation computations from continuously collected data. Efficient data derivation has given rise to the need of handling dataflow processes inside the database engine. Fig. 3 illustrates "In-DB ETL" where the dataflow is controlled by SQL query processing and computation is performed by UDFs.

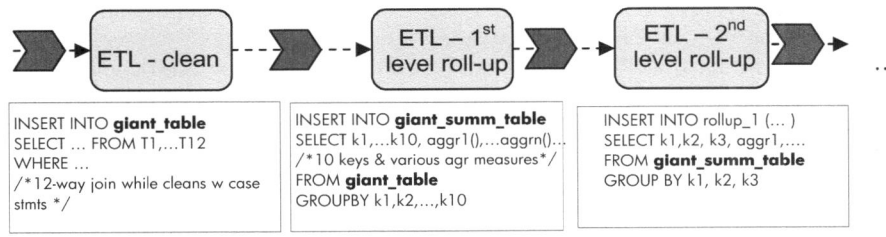

Fig. 3. In-DB ETL with dataflow by SQL and computation by UDFs

Fig. 4 shows another example where hydrographic information is monitored for predicting the status of river drainage networks. Under the conventional practice, applications manage models, compute derived information, and adapt or rebuild models separately from data management layer. As a result, large quantity of data travels from DB system to application and back, the applications are hard to scale, to parallelize, to manage and to trace.

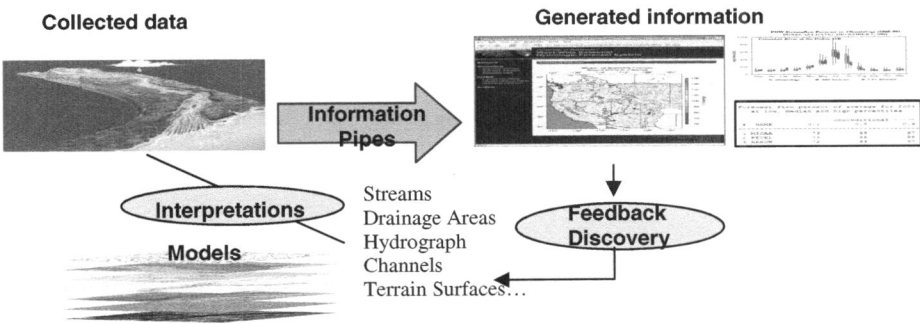

Fig. 4. A Hydrographic information flow and derivation example

However, a single SQL statement has limited expressive power at the application level, since the data flow represented in a SQL query is coincident with the control flow, but an application often requires additional data flows between its steps. In the other words, the query result at a step cannot be routed to more than one destination, while the requirements of the dataflow of an application are often modeled as a DAG (Directed Acyclic Graph).

The general solution we take is to represent correlated data flows by multiple SQL statements which are linked by RVFs. As mentioned before, RVFs have the same signatures as relational operators.

3.2 Correlated Query Process with RVF

A SQL statement expresses the composition of several data access and manipulation functions. A query execution plan can be viewed as a process including sequential and parallel steps, which opens the potential of handling queries at the process level.

For example, Pig Latin developed at Yahoo Research combines the high-level declarative querying in the spirit of SQL, and low-level, procedural programming. We share the same view as Pig Latin in treating a query as a process; however, beyond such a view, we have our specific research goals. As a database centric solution, we specify "steps" as individual SQL queries to be optimized by the underlying database engines. For modeling complex applications, we consider multiple correlated queries together with complex data flows and control flows represented as a DAG rather than a single query tree.

At the process level, multiple correlated SQL queries and RVFs form the sequential or concurrent steps of that application with complex data flows between them. Each step is an individual query or RVF that results in a row set.

In a regular query, the data flow and control flow are consistent and represented by a query tree. Even when there is a nested structure, a single query is unable to represent a general DAG-like data flow and control flow. Correlating multiple queries (including RVFs) into a process allows us to express control flows separately from data flows.

Refer to Fig. 5, for instance, an application is modeled as a query Q, followed by RVF f that takes Q's results as input, then followed by RVFs g_1 and g_2 which take f's as well as Q's results as input. The data flows and control flows of this application are not coincident. In order to express data flows separately from control flows, and to ensure the involved query Q and RVF f to be executed only once, this application cannot be expressed by a single SQL statement, but by a list of correlated queries at the process level. Conceptually the data dependency in the above example can be expressed as a sequence $<Q, f, g_1, g_2>$ meaning that Q should be provided before f, ...etc; this data dependency sequence is not unique ($<Q, f, g_2, g_1>$ is another one) but correct. The control flow can be expressed by $[Q, f, [g_1, g_2]]$ where $[g_1, g_2]$ can be executed in parallel.

Generally, a *Query Process* is made of one or more correlated SQL queries, referred to as *query steps*, which may be sequential, concurrent or nested (for simplicity we omit certain details such as conditional branch). A query step is a query, which may invoke RVFs or other UDFs. A QP represents a data intensive application at the process level where data flows are allowed to be separated from control flows,

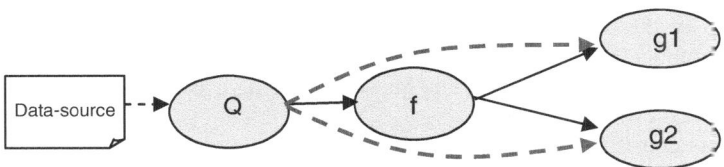

Fig. 5. A simple correlated query process, where data flows (solid and dash lines) and control flows (solid lines only) are not all coincident

and multiple start and exit points are allowed. We will illustrate the specification of a QP with an example in the next subsection.

3.3 Data Continuous Query Process

Motivated by automating enterprise information derivation processes, we have introduced a new kind of dataflow process - Data-Continuous Query Process (DCQP), which is data-stream driven and continuously running.

The basic operators of a DCQP are queries and RVFs which can be triggered repeatedly by *stream inputs*, *timers* or *event-conditions*. The sequence of executions generates a data *stream*.

To capture such data continuation semantics we have introduced the notion of *station* for hosting a continuously-executed queries and RVFs:

- A station is a named entity which hosts a query.
- When two stations are consecutive in the dataflow, a *pipe* from the upstream station, say S1, to the downstream station, say S2, is implied in the DCQP where the query hosted in S2 references S1 as (one of its) input, i.e., S1's output becomes input to S2.
- If a station S1 has multiple successor stations, then (conceptually) its output is to be replicated and sent to all its successor stations as input (fork).
- If a station S has multiple predecessor stations, then S is eligible for execution when all its predecessors have produced output (join).

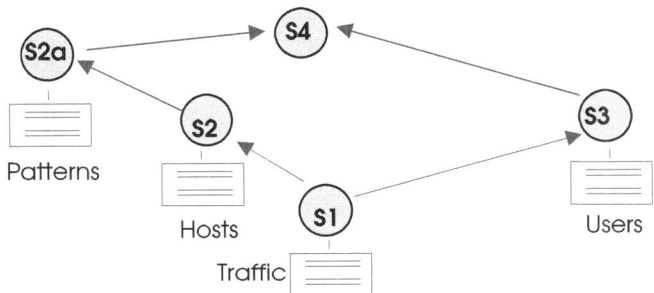

Fig. 6. A Query Process for network traffic analysis

Activated by the availability of input data, the presence of events or the timer, stations communicate asynchronously. A station is specified by its hosted query, and optionally with events and timers. The query specifying a station uses its predecessor stations as its data source.

A DCQP is constructed in the SQL framework and modeled as a graph of stations (nodes) and links between them. A simplified DCQP example for network traffic analysis is given in Fig. 6 where station S_1 captures point to point network traffic volumes, S_2 summarizes the host-to-host traffic volume; S_{2a}, an RVF, derives the host-to-host traffic pattern; S_3 gets user-to-user traffic; and S_4 as another RVF analyzes the patterns. The pseudo specification of this DCQP is illustrated below.

```
Create QP Traffic_Analysis {
  Source: Traffic STREAM, Hosts, Patterns, Users

  Define Station S1 As
    SELECT from-ip, to-ip, SUM(bytes) AS bytes   FROM Traffic GROUP BY from-ip, to-ip;
    Trigger: time interval

  Define Station S2 As
    SELECT h1.host-id AS f-host, h2.host-id AS t-host, S1.bytes FROM S1, Hosts h1, Hosts h2
    WHERE h1.ip = S1.from-ip AND h2.ip = S1.to-ip;

  Define Station S2a As
    SELECT * FROM Assign_pattern (S2, Pattern);

  Define Station S3 As
    SELECT u1.user-id AS from-user, u2.user-id AS to-user, S1.bytes FROM S1, Users u1,
    Users u2 WHERE u1.ip = S1.from-ip AND u2.ip = S1.to-ip;

  Define Station S4 As
    SELECT * FROM Pattern_analysis (S2a, S3);
}
```

3.4 Implementation Issues

We push down the Query Process support to the parallel database engine for fast data access and reduced dataflow, and in this way turn the database engine into a parallel computing engine as well. Layered on top of the parallel data management infrastructure, the QP manager can directly inherit the built-in data parallelism and distributed processing capability, automatically support partitioned parallelism, pipelined parallelism, subquery parallelism, etc.

In addition, we support query process execution with the following features:

- The query process definition is converted to a set of SQL queries hosted at stations.
- A station is provided with data caching capability.
- As the resulting data may be sent to multiple successor stations, an abstract notion of pipe is introduced to connect two stations; subsequently a pipe is realized as a queue, and there exist multiple ways to implement queues. We primarily focus on memory-based queues with support for overflow.

Although the ultimate goal is to extend the query executor of a parallel database engine to be "process aware", our experience shows that query processes can be realized to a great extent within the existing query executor where RVF support is added., For example, the following query

```
SELECT x, SUM(y) FROM rvf("SELECT a, b, FROM R1", "SELECT c, d FROM R2")
    GROUP BY x;
```

expresses three queries where the two "parameter queries" of *rvf* are wrapped by the outer-most query that takes the result of *rvf* execution as data source. We have supported the argument evaluation of an RVF's relation input with individual queries, where the RVF acts as the consumer of these query results and applies additional computation. Under this theme and with the pipe mechanism mentioned above, a QP can be executed by the query executor naturally.

4 Conclusions

By recognizing the scale, data bandwidth, and manageability challenges of ETL & BI analytics applications, we propose to leverage parallel processing capability in parallel data management, generalize and extend with declarative data flows and user-defined computations, to create a highly data-parallel analytics layer inside the database engine. This approach can significantly improve the end-to-end scale of BI systems, enabling very large scale operational BI.

Introducing RVFs is essential for constructing Query Processes for general-purpose dataflow-oriented computation. An RVF can model more complicated applications than scalar, aggregate, and table functions; can use multiple relations as input, and returns a relation as output; and appears wherever a relation list is expected, so they are easily composed in SQL.

We have built an initial prototype on a proprietary parallel database engine while experiments with an open-sourced database engine are also on-going. Our experience reveals the scalability, efficiency and flexibility of the proposed approach.

References

1. Chen, Q., Hsu, M.: Data-Continuous SQL Process Model. In: Proc. 16th Int. Conf. CoopIS 2008 (2008)
2. Dean, J.: Experiences with MapReduce, an abstraction for large-scale computation. In: International Conference on Parallel Architecture and Compilation Techniques. ACM, New York (2006)
3. DeWitt, D., Stonebraker, M.: MapReduce: A major step backward. The Database Column (2008), http://www.databasecolumn.com/2008/01/mapreduce-a-major-step-back.html
4. Gray, J., Liu, D.T., Nieto-Santisteban, M.A., Szalay, A.S., Heber, G., DeWitt, D.: Scientific Data Management in the Coming Decade. SIGMOD Record 34(4) (2005)
5. Greenplum, Greenplum MapReduce for the Petabytes Database (2008), http://www.greenplum.com/resources/MapReduce/
6. Isard, M., Budiu, M., Yu, Y., Birrell, A., Fetterly, D.: Dryad: Distributed data-parallel programs from sequential building blocks. In: EuroSys 2007 (March 2007)
7. Jaedicke, M., Mitschang, B.: User-Defined Table Operators: Enhancing Extensibility of ORDBMS. In: VLDB 1999 (1999)

Simplifying Information Integration: Object-Based Flow-of-Mappings Framework for Integration

Bogdan Alexe[1], Michael Gubanov[2], Mauricio A. Hernández[3], Howard Ho[3], Jen-Wei Huang[4], Yannis Katsis[5], Lucian Popa[3], Barna Saha[6], and Ioana Stanoi[3]

[1] University of California, Santa Cruz
[2] University of Washington
[3] IBM Almaden Research Center
[4] National Taiwan University
[5] University of California, San Diego
[6] University of Maryland

Abstract. The Clio project at IBM Almaden investigates foundational aspects of data transformation, with particular emphasis on the design and execution of schema mappings. We now use Clio as part of a broader data-flow framework in which mappings are just one component. These data-flows express complex transformations between several source and target schemas and require multiple mappings to be specified. This paper describes research issues we have encountered as we try to create and run these *mapping-based data-flows*. In particular, we describe how we use *Unified Famous Objects* (UFOs), a schema abstraction similar to business objects, as our data model, how we reason about flows of mappings over UFOs, and how we create and deploy transformations into different run-time engines.

Keywords: Schema Mappings, Schema Decomposition, Mapping Composition, Mapping Merge, Data Flows.

1 Introduction

The problem of transforming data between different schemas has been the focus of a significant amount of work both in the industrial and in the research sector. This problem arises in many different contexts, such as in exchanging messages between independently created applications or integrating data from several heterogeneous data sources into a single global database.

Clio [14,16,9], a research prototype jointly developed by IBM Almaden and the University of Toronto, investigated algorithmic and foundational aspects of schema mappings. Figure 1 depicts Clio's architecture and demonstrates, at a high-level, the steps involved in transforming data structured under a schema S (called the *source schema*) to data structured under a different schema T

M. Castellanos, U. Dayal, and T. Sellis (Eds.): BIRTE 2008, LNBIP 27, pp. 108–121, 2009.

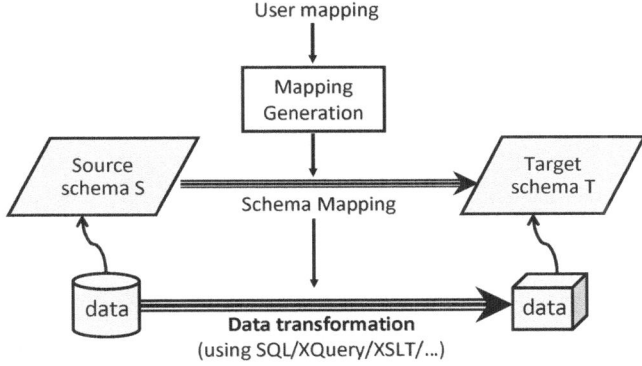

Fig. 1. Clio Architecture

(called the *target schema*). First, the user specifies a set of correspondences between attributes of S and T using Clio's graphical user interface. Based on these
attribute-to-attribute correspondences and the semantic constraints expressed
by S and T, Clio generates a declarative specification of the data transformation. This declarative specification, which we call *schema mapping*, is formally
expressed as a set of logical *source-to-target constraints* [8]. Finally, Clio translates the declarative schema mapping to executable code that performs the data
transformation.

Clio contains several code generation modules that target different languages.
For instance, when mapping xml-to-xml data, Clio can generate XQuery and
XSLT scripts from the same mapping. Alternatively, when mapping relational
data, Clio can generate SQL queries.

Complex information integration tasks generally need the orchestration or
flow of multiple data tranformation tasks. For instance, data transformations
commonly contain many intermediate steps, such as shredding nested data into
relations, performing joins, computing unions or eliminating duplicates. Parts of
the flow are in charge of extracting relevant information from raw data sources,
while other parts are in charge of transforming data into a common representation. Later parts of the flow are in charge of deploying the trasformed data into
target data marts or reports. Single monolithic mappings, those that map from a
source schema into a target schema, cannot capture these complex tranformation
semantics. And even in cases when a mapping can capture the tranformation semantics, the source and target schemas might be large and complex, making it
difficult for designers to create and maintain the mapping.

In this paper we discuss *Calliope*, a data flow system in development at IBM
Almaden that uses mappings as a fundamental building block. Instead of designing a monolithic schema mapping, the users of *Calliope* create a flow of smaller,
relatively simpler and easier to understand, mappings among small schemas. As
the name implies, the mappings are staged in a dependency graph and earlier
mappings produce intermediate results that are used by later mappings in the

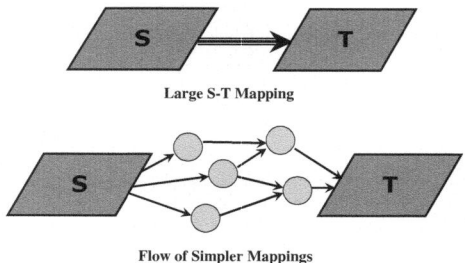

Fig. 2. Monolithic mappings (traditional approach) and flow of mappings in *Calliope*

flow (Figure 2). The individual mappings themselves can be designed using a mapping tool (e.g., such as Clio, which is now a component of *Calliope*). We believe the adoption of flows of mappings improves reusability, as commonly used transformation components can be reused either within the same transformation task (e.g. if an address construct appears twice within the same source schema) or across transformation tasks.

However, representing transformations as flows of mappings creates a number of interesting technical challenges. First, we need a uniform data model that represents data flowing over the graph of mappings. Second, to better understand the transformation semantics of a flow of mappings, we need to compose and *merge* mappings before generating run-time objects. Last, we need to generate transformation scripts over a combination of different runtime systems, each with different data transformation capabilities. We now briefly discuss each of these challenges in the rest of this section.

Unified Famous Objects (UFOs). One of the goals of *Calliope* is to raise the level of abstraction from schemas, and bring it one step closer to business objects. This is accomplished through the use of *Unified Famous Objects (UFOs)*. A UFO is a unified representation of a high-level concept, such as employee or address. One or more of these concepts may be embedded in a schema. Instead of mapping directly from an arbitrary complex source schema to an arbitrary complex target schema, which requires understanding their exact relationship, the mapping designer can focus on a single schema fragment at a time. Each such fragment is mapped to one or more UFOs that capture the concepts represented in the current schema fragment. The specification of the end-to-end transformation is then assembled from the smaller mappings between the source schema and the UFOs, the mappings between the UFOs and the target schema as well as from mappings between the UFOs themselves. We believe this approach facilitates thinking in terms of high-level concepts, which are closer to what users understand. UFOs also allow mapping reuse. Mappings between UFOs can be reused in various contexts where transformations are needed for the concepts they represent. Finally, this approach is modular: When changes occur over the schemas, or the semantics of the desired transformation changes, only the relevant parts of the smaller mappings involving UFOs need to be updated. In this new approach the mapping designer will be able to focus on the mappings between the end schemas and the UFOs, and reuse as much as possible previously designed mappings between UFOs.

Mapping Merge. To allow the execution of flows of mappings, *Calliope* employs novel mapping technology that allows for the automatic assembly of initially uncorrelated mappings into larger and richer mappings. This technology relies on two important mechanisms: mapping merging and mapping composition. *Mapping merging*, or correlation, is responsible for joining and fusing data coming from initially uncorrelated mappings. *Mapping composition* [7,2] on the other hand, allows assembling sequences of mappings into larger mappings. Intuitively, mapping merging combines mappings that correspond to parallel branches in the mapping flow graph, while mapping composition combines mappings that correspond to sequential paths. We note that mapping merging, which is discussed in some level of detail in Section 4, is an operator on schema mappings that has been largely undeveloped until now.

Unified Flow Model (UFM) Framework. Some of our use cases involve users that want to design a single transformation using a *combination* of many different data transformation tools, such as mapping tools, ETL tools or query languages [3]. Such cases are becoming increasingly common in practice for two main reasons: First, some data transformation tools are not expressive enough to represent a data transformation and thus the transformation generated by the tool has to be augmented with additional operators. For instance, Clio does not allow sorting of results. Therefore if sorting is desirable, the transformation generated by Clio has to be augmented with sorting in a language that offers the appropriate construct, such as SQL. Second, even when a single data transformation tool can express the entire transformation, users familiar with different tools want to collaborate on the design of the same transformation. For example, analysts designing the coarse outline of a transformation in Clio would like to have it extended with lower-level details by programmers, who are familiar with ETL-tools.

To address these interoperability requirements we designed the *Unified Flow Model (UFM) framework*. At the heart of this framework lies the Unified Flow Model (UFM), which allows the representation of a data transformation in a tool-independent way. Given the UFM framework, all it takes to make *Calliope* interoperate with other data transformation tools is to design procedures that translate the internal representation of any data transformation tool to UFM and vice versa. The main challenges in the context of this framework is finding the right language for UFM and designing the translations between UFM and the internal languages for the various data trasformation tools.

The following sections describe in more details the main components of *Calliope*. Section 2 introduces the notion of UFOs and describes the UFO repository. Section 3 describes how to decompose a schema into a set of UFOs. Section 4 gives an illustration of the mapping merging technique used in *Calliope*, and Section 5 presents the UFM framework.

We note that *Calliope* is at an early stage and many of the ideas presented in this paper are still under development.

2   Unified Famous Objects (UFOs)

Traditional mapping tools allow the specification of a transformation by defining a mapping from a source schema to a target schema. However this approach becomes problematic as the schemas become larger and the transformation more complex. Large schemas are hard to understand and it is even harder to create a mapping between them in a single step. To remedy this problem, *Calliope* allows users to split the large and complex mapping of the source to the target schema into more easily comprehensible steps (which are themselves composed of multiple mappings) that are based on the use of intermediate Unified Famous Objects (UFOs).

A UFO, similar to a business object, is a flat object representing a simple concept, such as an employee, a product or an article. Being similar to a business object, it is a higher-level abstraction than a schema and, as such, closer to the understanding of the mapping designer. In *Calliope*, UFOs can be either defined manually or extracted automatically from a source that provides standardized schemas, such as Freebase or OAGI. Once created, they are stored in the metadata repository of the system, ready to be used in mappings. To model semantic relationships between related UFOs, the metadata repository can also hold mappings between UFOs.

Given the metadata repository, mapping a source schema to a target schema translates to the following steps: First, the designer finds the UFOs in the metadata repository that are relevant to the source schema. To facilitate this process, the metadata repository offers an interface that allows the designer to browse and query its contents. Once the relevant UFOs are found, the designer creates a separate mapping between the source schema fragment representing a particular concept and the corresponding UFO. After finishing this process for the source schema, the designer repeats the symmetrical procedure for the target schema: find UFOs that are relevant to the target schema and design mappings from each of them to the target schema. The resulting end-to-end transformation between the source schema and the target schema is then the flow of mappings composed of: a) mappings from the source schema to UFOs, b) mappings from UFOs to the target schema and c) any number of intermediate mappings that may be needed between the UFOs themselves. Some of the intermediate UFO-to-UFO mappings may have to be designed at this point, but some could be reused from the metadata repository (if they already exist).

Figure 3 shows a sample flow of mappings between source and target schemas in the presence of UFOs. The picture displays two types of nodes: schema nodes (used to represent both source/target schemas and UFOs) and mapping nodes. Source schemas (inside the box on the left) are mapped to UFOs (inside the box in the middle). For instance, the source on the top contains information on projects and employees and therefore it has been mapped to the corresponding two UFOs, representing projects and employees, respectively. Apart from

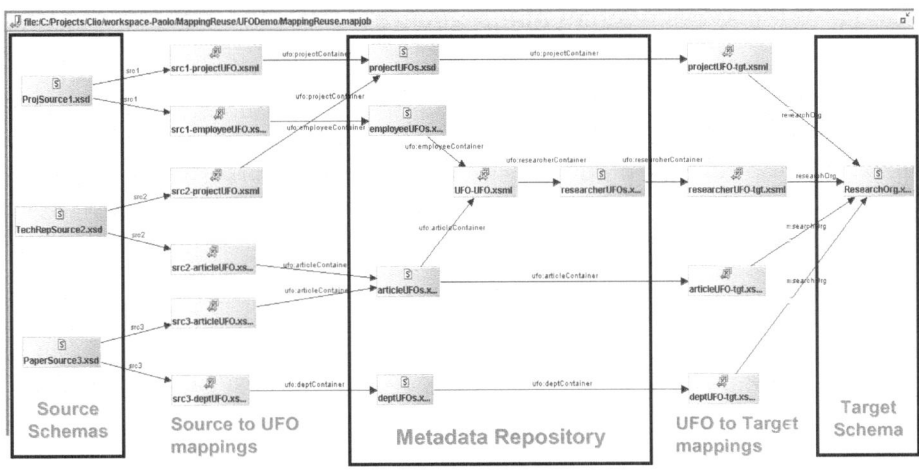

Fig. 3. Flows of Schema Mappings

the mappings from source schemas to UFOs, the picture also shows mappings between UFOs (inside the middle box). For example, the UFO representing a researcher contains both employee and article information and therefore it has mappings from the corresponding UFOs. Finally, UFOs are mapped to the target schema (shown inside the box on the right).

Using UFOs in the mapping process yields several advantages: First, it allows the designer to decompose a large source-to-target mapping into many smaller mappings that involve the UFOs. Since each individual mapping is relatively small, it is easy to create and maintain. Second, by storing in the repository the mappings between schemas and UFOs, we can improve the precision of matching algorithms by learning from previous mappings. For instance, we can store attribute name synonyms next to each UFO attribute to help with subsequent matching. More importantly, UFOs allow for standardization and reuse. A large part of the mapping and transformation logic can be expressed in terms of a fixed set of UFOs describing a domain, and this logic can then be reused and instantiated in different applications (on the same domain).

The presence of the UFO repository also creates some important challenges. First, as the number of UFOs increases it becomes increasingly harder to find the UFOs that are relevant to a given schema. To remedy this problem, *Calliope* contains a schema decomposition algorithm, described in the following section. Second, the mappings that are created between the source and target schemas and the UFOs, as well as the mappings that relate UFOs and are potentially extracted from the repository, are initially uncorrelated, and possibly independently designed. The main challenge here is to orchestrate the flow of uncorrelated mappings into a global mapping that describes a meaningful end-to-end transformation. As a solution to this problem, *Calliope* relies on a mapping merge mechanism. We give an overview of this mechanism in Section 4.

3   Schema Decomposition

Today a user will handle new schemas integration tasks by designing the transformation operator flow manually, from scratch. Since defining these operators is cumbersome [14,16], it is important to bootstrap the process with relevant operator flows. At the core of reusing operator flows is the task of recognizing commonalities between a new input schema and fragments of previously used schemas. Thus, an essential step in using UFOs for schema mapping is the *decomposition* of the source and target scheams into the right set of UFOs. *Schema decomposition* is the technique of automatically selecting a collection of concepts from a given repository, which together form a good coverage of a new input schema. In general, schema decomposition enables the understanding and representation of large schemas in terms of the granular concepts they represent. This step should automatically propose good decompositions, and allow the user to further modify and enhance them. Schema decomposition should be both efficient in tranversing a large space of UFOs, and effective in producing decompositions of high quality.

Consider for example the schema in Figure 4(i). An integration advisor, using the techniques of schema decomposition can automatically identify that the same *Address* format has been used before. It will then propose transformation flows for this specific format of *Address*.

There has been a lot of work in semi-automated schema matching, proposing solutions that are based on schema and instance level, element and structure-level, language and constraint-based matching, as well as composite approaches [17,18]. Some of the different systems that have been developed for matching include SemInt [10], CUPID [12], SF [13], LSD [5], GLUE [6], COMA [4], TranScm [15], Momis [1] etc. The reuse of previously determined match results proposed

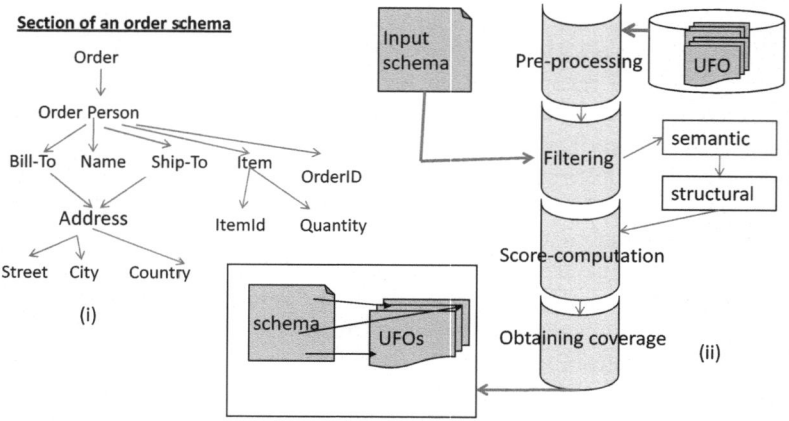

Fig. 4. (i) A Schema-Graph, (ii) Components Used in Schema Decomposition

by Rahm and Bernstein in [17] has also attracted recent research focus and has been successful in significantly reducing the manual match work [4,11]. However the reuse of matching is limited to element to element match and not on any larger matching concepts. Most of these works can only handle small and structurally simple schemas and define direct schema matching from source to target. In contrast, our proposed schema decomposition technique can be used to exploit reuse on much larger matching concepts (*UFOs*) and make the task of schema-matching and mapping between structurally complex schemas substantially easier.

Enabling reuse goes beyond the design of schema mappings. It can be exploited by ETL tools, Mashups and in general by Information Integration systems. Consider a schema for a new Mashup feed, where similar fragments of the schema have been used in other data sources. Then the automated advisor can detect similarities, and advise the user of opportunities for exploiting the new data in interesting ways. In Information Integration, let us consider two schemas that are understood in terms of the concepts from the repository. Then common concepts point to schema matches, and can be used to efficiently direct users through the task of integrating the two schemas. In all these aspects, schema decomposition will serve as the fundamental tool.

3.1   An Overview of the Technique

Given a source and a target schema, and a repository of *UFOs*, schema decomposition aims towards "covering" the schemas with UFOs as well as possible. Known matching and transformations between UFOs can then be exploited to obtain a source to target schema-mapping. Populating the repository with the right schema fragments can be thought of as a preprocessing step. By contrast, schema decomposition needs to be handled online, as new input schemas are introduced. Let us denote an input schema by S, and the repository by $\mathcal{R}$. Schema decomposition identifies the matching concepts (schema fragments) U_i's from $\mathcal{R}$ that best cover S. The selection of the U_i's and their corresponding positioning to cover S are factors in the quality of the coverage. For efficiency, we use a filter/evaluate/refine approach as described below.

The repository of schema fragments U_i can be quite large, and most of these fragments will not match at all S. Therefore, it is necessary to reduce the search space and select only fragments from the repository that have potential to be relevant. This filtering is performed in two phases. In the first phase a semantic filter is used that compares labels of each U_i against labels in S. Note that the structural relationship between labels is not yet taken into account. The output of the semantic filter is the set of candidates U_i that have any semantic labels in common with S. However, this does not test for the structural similarity. That is, do the labels in any of the U_i's match a cohesive region of S? And if yes, how much structural flexibility do we allow when considering a match? The structural filter evaluates these questions and further refines the candidate set.

The next step in schema decomposition involves a score computation model, in order to compute the quality scores for the U_i's in the candidate set. The score for each U_i is computed based on the portions of S that U_i can potentially cover. Note that considering all possible sub-regions of S and calculating quality scores of U_i's for each region is an exponential process. In practice the score computation model needs to rely on a quality measure that avoids this exponential running time and can be computed efficiently.

The last step is an optimization step that considers all the selected *UFOs* together with their quality score and 1) selects a subset of *UFOs* and 2) positions them relative to S so that the aggregated quality score is maximized. Figure 4(ii) illustrates this framework. This last step is NP-Complete and thus an exhaustive approach will take exponential time and is potentially of no practical interest. Instead, approximation algorithms that are fast and guarantee near-optimality of the result quality should be used.

4   Orchestrating Flows of Schema Mappings

As mentioned before, the mappings between the end schemas and the UFOs, as well as the mappings that relate UFOs extracted from the repository, are initially uncorrelated. To express a meaningful end-to-end transformation, the flow of independently designed, uncorrelated mappings needs to be assembled into a global mapping. To achieve this, *Calliope* relies on a mechanism for merging smaller, uncorrelated mappings into larger, richer mappings, through joining and fusing data coming from individual mappings.

We give a brief overview of the merging technique through the use of a simple example. Consider the scenario in Figure 5. It consists of a nested source schema representing information about projects and the employees associated with each project. Potentially following the schema decomposition phase, a set of three UFOs have been identified as relevant to the source schema: Project, Dept, and Employee. These are possibly extracted from the UFO repository and are

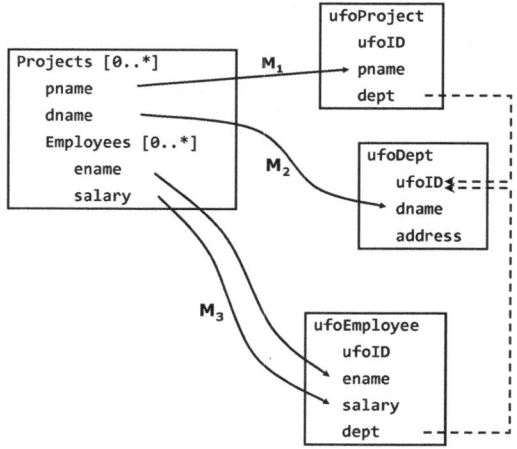

Fig. 5. Mapping Merge Scenario

standardized representations of the business objects project, department, and employee, respectively. Using schema matching techniques and the mapping design capabilities in Clio, three initial mappings M_1, M_2, M_3 are constructed between the source schema and each of the three UFOs. These mappings are decorrelated and, in a real life scenario, may be constructed by independent mapping designers. This mapping scenario may be part of a larger flow of mappings from the source schema, through the UFOs above, and possibly others, to one or more target schemas. The mappings in the scenario above can be expressed in a logical formalism similar to source-to-target constraints [8], as below:

$$M_1 : \text{Projects}(p, d, E) \rightarrow \text{ufoProject}(\text{PID}, p, \text{DID})$$
$$M_2 : \text{Projects}(p, d, E) \rightarrow \text{ufoDept}(\text{DID}', d, A)$$
$$M_3 : \text{Projects}(p, d, E) \wedge E(e, s) \rightarrow \text{ufoEmployee}(\text{EID}, e, s, \text{DID}'')$$

For instance, mapping M_3 above states that for each project record in the source, and each employee record in the set of employees associated with that project, there must exist a ufoEmployee object where the values for the ename and salary attributes come from the corresponding attributes in the source employee record. However, according to M_3, the identifier of the ufoEmployee object as well as the value of the dept attribute remain unspecified, and are allowed to take some arbitrary values. Similarly, mappings M_1 and M_2 put in correspondence source project records to ufoProject and ufoDept objects, respectively. Note that according to the mappings above, the values of the dept attribute in ufoProject and ufoEmployee and the identifier of ufoDept are not correlated.

The mapping merge mechanism in *Calliope* rewrites and combines the initial mappings to generate a set of richer mappings that support a meaningful transformation where target data records the "correct" correlations. As a first step, the merging technique takes advantage of any constraints that may be present among the UFOs. In the example above, there are two referential constraints, indicated through dashed lines in figure 5. These constraints indicate that the dept attributes in ufoProject and ufoEmployee must be identifiers for a ufoDept. The merge mechanism considers these constraints and rewrites the initial mappings into the following mappings:

$$M_1' : \text{Projects}(p, d, E) \rightarrow \text{ufoProject}(\text{PID}, p, \text{DID})$$
$$\wedge \text{ufoDept}(\text{DID}, d, A')$$
$$M_3' : \text{Projects}(p, d, E) \wedge E(e, s) \rightarrow \text{ufoEmployee}(\text{EID}, e, s, \text{DID}'')$$
$$\wedge \text{ufoDept}(\text{DID}'', d, A'')$$

In the new set of mappings, M_1' is the result of merging M_1 and M_2, and M_3' is the result of merging M_3 and M_2. Note that in M_1', the ufoProject and ufoDept are correlated via the department identifier field. A similar correlation exists in M_3' between ufoEmployee and ufoDept. Note that there is no correlation (yet) between M_1' and M_3'.

An additional merging step applies by taking advantage of the fact that M_3' is a "specialization" of M_1': the latter mapping defines behavior for project records, in general, whereas the former considers, additionally, the employee records associated to the project records. Concretely, we can merge the more specific part of M_3' (i.e., the employee mapping behavior) into M_1' as a nested sub-mapping. The resulting mapping is the following:

$$\text{Projects}(p, d, E) \rightarrow \text{ufoProject}(PID, p, DID)$$
$$\wedge \text{ufoDept}(DID, d, A')$$
$$\wedge [\ E(e, s) \rightarrow \text{ufoEmployee}(EID, e, s, DID)\ ]$$

The resulting overall mapping includes now all the "right" correlations between projects, departments and employees. The top-level mapping transforms all project records, by creating two correlated instances of `ufoProject` and `ufoDept` for each project record; the sub-mapping, additionally, maps all the employee records that are associated with a project, and creates instances of `ufoEmployee` that are all correlated to the same instance of `ufoDept`.

This simple example gives just an illustration of the techniques for merging mappings that are used in *Calliope*. The full details of the complete algorithm for merging will be given elsewhere.

5   Unified Flow Model Framework

As a schema mapping tool, Clio aimed in allowing people to transform data between different schemas. However, it is far from being the only tool available for this task. Data transformations can nowadays be carried out through a variety of different tools, such as mapping tools (e.g. Clio, Stylus Studio), Extract-Transform-Load (ETL) tools (e.g. Datastage), Database Management Systems (DBMSs) (e.g. DB2, SQL Server, Oracle) etc.

These tools offer different *paradigms to design* the transformation (and thus they have different target audiences) and they employ different *engines to execute* the designed transformation. Table 1 summarizes the design and runtime components for mapping tools, ETL tools and DBMSs. For instance, mapping tools allow the declarative design of transformations (through mappings) and therefore they are best suited to analysts who want to design a transformation at a high level. In contrast, ETL tools have a more procedural flavor, allowing the design of a transformation through the composition of a set of primitive ETL operators. This makes them the ideal platform for highly-skilled programmers who want to design complex workflows. Similarly, mapping tools and ETL tools employ different engines to execute the transformation. Mapping tools usually translate the transformation down to a query language like XQuery or XSLT, while ETL tools often contain their custom execution engines optimized for the supported set of operators.

However by tying together a certain design component and runtime component, existing tools severely restrict the users in two significant ways: First, they force the users to employ both the design and runtime component supported by the tool. For example, a user cannot design a transformation in Clio

Table 1. Properties of Mapping Tools, ETL Tools and DBMSs

	Mapping Tools	ETL Tools	DBMS
Design Paradigm	Mappings	Flow of ETL Operators	Query
Target Audience	Analysts	Programmers	DB Experts
Execution Engine	XQuery engine, XSLT engine etc.	Custom	Query Engine

and execute it using the highly optimized internal engine of an ETL tool, since the ETL's runtime engine can only execute transformations designed through the ETL paradigm. Second, they do not allow users to employ different design paradigms to design a single flow. For example, an ETL programmer willing to extend the transformation sketched by an analyst in Clio, has to start from scratch as there is no automatic provision for the translation between design paradigms supported by different tools.

To alleviate these problems, *Calliope* is based on the *Unified Flow Model Framework*, shown in Figure 6. The heart of this framework is the *Unified Flow Model* (UFM); a model that represents data transformations in a tool-independent way. Different design components can be added in the system in a modular way by specifying a procedure for translating the internal representation of the design component to UFM and vice versa. Similarly, a developer can add an execution engine to the framework by specifying how a transformation described in UFM can be translated to a language recognized by the particular execution engine. The result is a framework, where users can design a data transformation in one or more design components (originating from diverse tools, such as mapping tools, ETL tools etc.) and subsequently execute it on any execution engine that has been added to the framework.

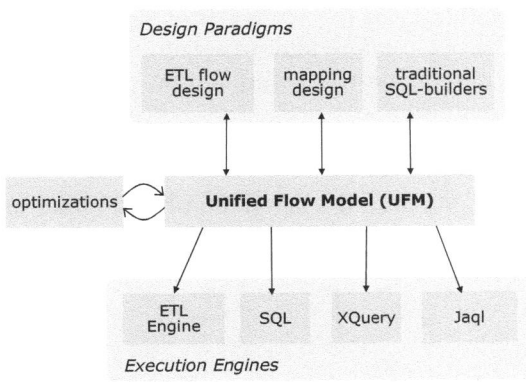

Fig. 6. Unified Flow Model Framework

The implementation of the UFM framework poses many important challenges caused by the differences between various design components (respectively execution engines). First of all, the translation between different design paradigms might not be always possible, since they have in general different expressive power. Should such translations fail or should the system try to translate the largest subset of the transformation possible? Second, although a transformation can be optimized at the UFM level, there will be some optimizations that are execution engine dependent. What types of optimizations can be done on the UFM level and which optimizations require knowledge of the execution engine on which the transformation is going to be ran? These are a few of the questions that arise in the implementation of the UFM framework.

As a first step towards our vision, we have designed the following components of the UFM framework: a) the UFM model as a flow of ETL-like operators that can express most common data transformations (which contain all transformations currently supported by Clio's mapping language), b) the translation from Clio's mapping language to UFM and c) the translation from UFM to Jaql; a JSON query language that contains a rewriting component that translates queries to MapReduce jobs that can be executed in Hadoop. This corresponds to the addition of two components in the architecture shown in Figure 6: the mapping design component (with a unidirectional arrow into UFM) and the Hadoop execution engine.

6 Conclusion

We have discussed the main components of *Calliope*, a system for creating and maintaining flows of mappings. Our main motivation for *Calliope* was to extend and reuse the basic schema mapping operations explored in Clio in more complex and modular data transformation jobs.

We believe that complex mappings are difficult to build and manage in one step. Not only is it conceptually hard to understand the relationships between schemas with a potentially very large number of elements but it is also hard to visualize and debug them. *Calliope* allows users to express mappings in terms of higher-level objects, such as business objects, that are easier to understand and do not couple implementation with semantics. Further, mappings in *Calliope* are smaller and modular, increasing the opportunity for reuse.

Acknowledgements. We acknowledge Hamid Pirahesh for his original suggestion to use UFOs and also for his continuous feedback on this work.

References

1. Bergamaschi, S., Castano, S., Vincini, M., Beneventano, D.: Semantic integration of heterogeneous information sources. Data Knowl. Eng. 36(3), 215–249 (2001)
2. Bernstein, P.A., Green, T.J., Melnik, S., Nash, A.: Implementing Mapping Composition. In: Proceedings of VLDB, pp. 55–66 (2006)

3. Dessloch, S., Hernández, M.A., Wisnesky, R., Radwan, A., Zhou, J.: Orchid: Integrating Schema Mapping and ETL. In: ICDE, pp. 1307–1316 (2008)
4. Do, H.-H., Rahm, E.: Coma: a system for flexible combination of schema matching approaches. In: VLDB 2002, pp. 610–621 (2002)
5. Doan, A., Domingos, P., Halevy, A.Y.: Reconciling schemas of disparate data sources: a machine-learning approach. In: SIGMOD 2001, pp. 509–520 (2001)
6. Doan, A., Madhavan, J., Domingos, P., Halevy, A.: Learning to map between ontologies on the semantic web. In: WWW 2002, pp. 662–673 (2002)
7. Fagin, R., Kolaitis, P., Popa, L., Tan, W.-C.: Composing Schema Mappings: Second-Order Dependencies to the Rescue. In: PODS, pp. 83–94 (2004)
8. Fagin, R., Kolaitis, P.G., Miller, R.J., Popa, L.: Data exchange: semantics and query answering. Theoretical Computer Science 336(1), 89–124 (2005)
9. Fuxman, A., Hernández, M.A., Ho, H., Miller, R.J., Papotti, P., Popa, L.: Nested Mappings: Schema Mapping Reloaded. In: Proceedings of VLDB, pp. 67–78 (2006)
10. Li, W.-S., Clifton, C.: Semint: a tool for identifying attribute correspondences in heterogeneous databases using neural networks. Data Knowl. Eng. 33(1), 49–84 (2000)
11. Madhavan, J., Bernstein, P.A., Doan, A., Halevy, A.: Corpus-based schema matching. In: ICDE 2005, pp. 57–68 (2005)
12. Madhavan, J., Bernstein, P.A., Rahm, E.: Generic schema matching with cupid. In: VLDB 2001, pp. 49–58 (2001)
13. Melnik, S., Garcia-Molina, H., Rahm, E.: Similarity flooding: A versatile graph matching algorithm. In: ICDE 2002, pp. 117–128 (2002)
14. Miller, R.J., Haas, L.M., Hernández, M.A.: Schema mapping as query discovery. In: VLDB 2000, pp. 77–88 (2000)
15. Milo, T., Zohar, S.: Using schema matching to simplify heterogeneous data translation. In: VLDB 1998, pp. 122–133 (1998)
16. Popa, L., Velegrakis, Y., Hernández, M.A., Miller, R.J., Fagin, R.: Translating web data. In: VLDB 2002, pp. 598–609 (2002)
17. Rahm, E., Bernstein, P.A.: A survey of approaches to automatic schema matching. The VLDB Journal 10(4), 334–350 (2001)
18. Shvaiko, P., Euzenat, J.: A survey of schema-based matching approaches. In: Spaccapietra, S. (ed.) Journal on Data Semantics IV. LNCS, vol. 3730, pp. 146–171. Springer, Heidelberg (2005)

Author Index